Dedications

To our families – Amy, Amir, Bev, Doug, Virginia, George, Anita, Chris, James,
Jo and Albert for your enduring faith in us, your love, support and friendship.
Without you all, none if this would be possible.

Christian Topf – For your creativity, talent and patience.
Leslie and Jill Bishop – For believing in us. Thank you so much for all your support.
Tim Reith – For continually inspiring us through your friendship and enthusiasm.
Clare Marsh – For helping us create our home in Cornwall.
Matt Robinson – For introducing us to cob in England and for inspiring
us to start our journey in Cornwall.
Linda Smiley and Ianto Evans – Thank you for helping us to
open our eyes and for starting us off on this journey.

PHOTO CREDITS All photos are by Katy Bryce, except the following:
Ray Main (www.raymain.co.uk) – front cover middle left, bottom middle, bottom right, p.1, p.19, p.22 bottom right, p.24, p.25 left and right, p.59 bottom left, p.95 right, p.105 left, p.107 middle, p.109 right, p.116 left, p.118 bottom right, p.127, p.176, p.181 left, p.229, back cover bottom left; *Grand Designs Magazine* – back cover portrait of Adam and Katy; Jim Clayton – back top left, p.177 right, p.191, p.194 all photos, p.197 all photos; UNESCO – p.3 left and right, p.5, p.10, p.11 left, p.12, p.14 all except bottom right, p.17 all photos, p.108 left, p.151, p.177 left; Barbara Tremaine – p.32 left; Eco Arc Architecture/Green Oak Carpentry Company, Chithurst Buddhist Monastery – p.38, p.104, p.217 left and right; Clare Marsh – p.51 middle, p.65, p.178; Paul Finbow – p.72 bottom; Carpenter Oak – p.105 right, p.107 top left and right, middle left and middle; Ian Armstrong, ARCO$_2$ Architecture – p.115 top middle, p.128 middle left; Second Nature UK Ltd – p.123 middle and right, p.128 top and middle middle; Excel Industries – p.128 bottom middle; Christian Topf – front cover middle middle, p.22 bottom left and middle, p.35, p.40, p.47 top right, p.54 left and right, p.55 bottom left, p.115 bottom left; Associated Architects, Kate Ellis – p.25 middle; Anita Bryce – p.31 left, p.117 middle; Chris Bryce – p.138 left; James Bryce – p.3 middle, p.26, p.59 top right, p.79, p.98 left; Louise Cooke – p.11 right, p.115 top right; Beaford Arts Centre, Devon – p.22 middle left; Seven Generations Natural Builders – p.31 right, p.98 right, p.115 bottom right, p.181 right; Peter Harris – p.199 right; Rhodda Lloyd Travers Architects – p.217 middle, p.222.

Published in 2006 by Green Books Ltd, Foxhole, Dartington, Totnes, Devon TQ9 6EB
edit@greenbooks.co.uk www.greenbooks.co.uk
© Adam Weismann and Katy Bryce 2006 www.cobincornwall.com
Section header illustrations: Carl Homstad Pen & ink illustrations: Christian Topf
Design & layout: Christian Topf Design (CTD) www.ctd-studio.co.uk
All rights reserved. Printed by Butler & Tanner, Frome, Somerset, UK
Text & cover materials: Revive 100% recycled paper

ISBN 1 903998 72 7

DISCLAIMER: The authors and publishers accept no liability for personal injury, property damage or loss as a result of actions inspired by this book. Building work can be dangerous, and due care should always be taken.

building with cob

a step-by-step guide

Adam Weismann & Katy Bryce

with selected photographs by Ray Main

Contents

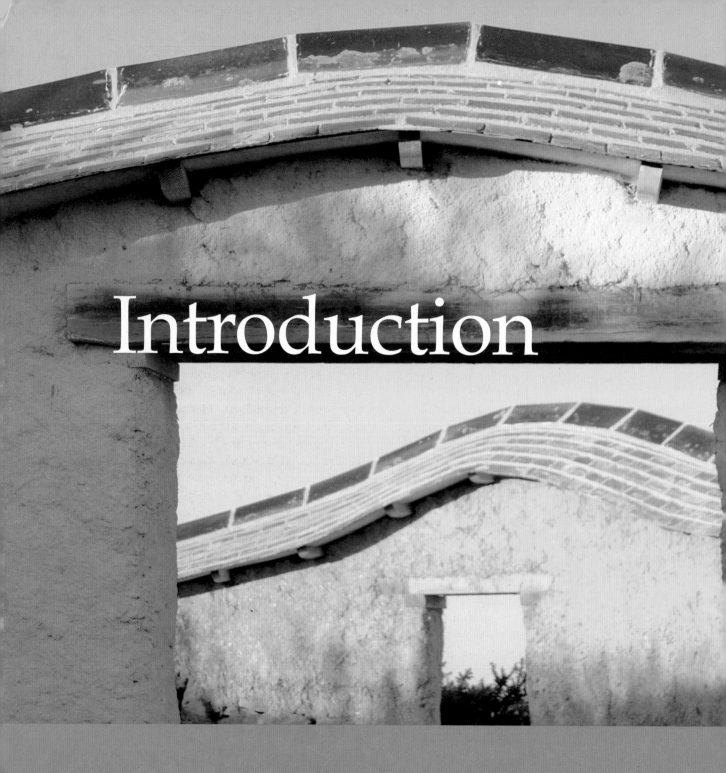

Introduction

This book is not about going off to live in a cave ... It is not based on the idea that everyone can find an acre in the country, or upon a sentimental attachment to the past. It is rather about finding a new and necessary balance in our lives between what can be done by hand, and what still must be done by machine. Lloyd Kahn – *Shelter*

We live in interesting times. The last 150 years, since the industrial revolution began and the technological age took root, have brought rapid, dramatic changes to the world we inhabit. Some of these changes could be said to have brought about vast improvements in the general conditions of most people's lives. On the other hand, some of these changes have brought about a false sense of progress, and two main outcomes have arisen. Firstly, the ecology of the planet is suffering badly. Although it is a naturally self-adjusting mechanism, and is designed to accommodate wastes and pollution, the changes that we have brought about have happened too quickly for it to adjust, and our levels of pollution and waste have become too much and too toxic for it to cope. We are at a crossroads. We can go one of two ways: either stumble blindly on into the future, and hope that something works itself out; or stop now, and start to make conscious changes on a personal level.

We can become aware and conscious of the small and large decisions that we make on a daily basis. One of the most fundamental decisions we can make is what sort of a house we live in. What sort of materials is it made out of? Are they local, renewable, non-toxic, requiring little energy to produce? Does the overall design of the house require little energy to heat and cool? Can it make

use of the free energy of the sun, and deal efficiently with wastes? Can it encourage communities to come together and build? Can it help to take some of the burden off our already stretched planet?

The second outcome that has arisen out of these dramatic and rapid changes has been our estrangement from the natural world. For without this estrangement, how could we have so easily and flippantly used and abused it so much? As we no longer directly relate to the natural world for our survival, for food, for shelter, there seems to have emerged a separation between wildlife and wilderness areas, and the 'civilised' world – shopping complexes and cityscapes. This is sad, not just because the natural world is suffering at the hands of our insensitivity (which will cause us suffering in the future, as it fights back), but because we too are suffering, through our estrangement to what is in fact a part of us, the whole of us. It should be a rich and rewarding symbiotic relationship: you give me some rocks and mud to build my house, and I'll do my best to honour and care for the land on which I am building.

To help us move forward, we can take a glimpse back to a period before the

industrial revolution – the last era in history when many people lived through this intimate, reciprocal relationship with the natural world. Or we can look at the tribes and communities throughout the world where industrialisation has not reached. We can study and learn from their buildings and dwellings, the ways they feed themselves, and their relationships with the land by which they are supported.

A unifying characteristic of these pre-industrial societies is their sense of holism, and their understanding that everything is linked, that all actions have an impact on all parts of the system, and that the whole is more important than the sum of its parts.

THE EVERLASTING CYCLE OF COB

EXTRACTING · MIXING · CONSTRUCTING · LIVING · DECAYING

To these traditional societies, progress is not seen, as it is in our societies, as a linear concept, moving along a straight line from the past into the present and into the future. In industrial societies, at each stage newer and more sophisticated things are invented, so that we feel that we are better off today than we were yesterday. In traditional societies, time is seen as a circle, ever linked to the eternal cycles of nature – the moon, the seasons, the spin of the earth around the sun, and the cycle of life and death.

4

Earth dwelling with hand-painted exterior decoration, India

On a metaphorical level, this book is about circles and cyclical time. On a practical level, this book is about building with cob (a simple mixture of clay subsoil, aggregate, straw and water). Through the process of building with cob we are encouraging you to reconnect to this wholesome, everlasting form. In the book we talk about coming full circle, back to those ideas and techniques of the past that really worked, such as walls built out of cob and roofs made out of thatch. We talk about the renewable cycles of natural building materials such as cob and stone – which the planet is constantly making and which can be eternally re-used; or thatch and sustainably harvested wood, which will decompose safely when they have reached the end of their life, go back into the earth whence they came, and be turned into something new to nourish the garden.

We talk also about the use of lime as a building material, which has its own cycle as it moves from the ground as limestone, is processed into a material that can be plastered onto walls, at which point it reacts with the air, and effectively turns back into limestone. Lime can also be removed from a building, re-mixed, and re-used again.

In the chapter on siting and designing a cob structure, we encourage you to become aware of the daily and monthly cycles of the sun, and to orient your building accordingly so that you can benefit from the light and heat it provides, and so that it will be comfortable and joyous to live in and experience through all the seasons. When we talk about building schedules, we encourage the builder to respect the seasonal rhythms of cob building, and to embark on a project during the best weather months – starting in spring, and completing in autumn.

There is also a natural cycle in the actual construction of a cob building. In an ideal situation, the materials that are present naturally on the site can be efficiently rearranged so that little waste is generated, costs are kept down, and transport of outside materials onto the site is kept to a minimum. For example, the turf from the potential building site can be removed, stored and later used to lay on top of a roof structure for a 'green' roof. The topsoil can be used to create flower or vegetable beds, and the subsoil, which has been removed to make way for the foundations, if suitable can be mixed with straw and aggregate to make cob to build the walls. You may also need to prune some surrounding trees to allow more sunlight in, and these prunings can be utilised in the roof structure. It is up to you how far you want to go.

It is emphasised that cob buildings may need more maintenance than a standard, modern house, such as a yearly lime washing. This again can reconnect us to the natural seasonal cycles of nature.

Consider also the information in Chapter 13 about cob and modern building regulations. Cob can be brought up to the standards required by the government in a modern dwelling through an approved, holistic method of assessment.

Most importantly, although cob can be moulded and formed into whatever shapes you desire, it is best and at its strongest when built in the round: efficient, because no heat can get lost in corners, and cosy as it encircles you in an eternal embrace.

Ultimately, building with cob and other natural materials is a way that we can literally get back in 'touch' with nature. Building with cob is about getting your hands dirty, touching the stuff, feeling its stickiness, its grittiness, its pliability and plasticity. We encourage people to wake up their senses, to learn to recognise the earth – to understand its suitability for cob building, not through rigorous, scientific tests (although these can be done), but through learning what it should look, smell, and taste like. In this way, although this book is practical and grounded in technical matters – building a solid structure is a very serious business – it also encourages the individual to re-engage with the art of building and have a lot of fun in the process.

Hopefully, once the techniques outlined in this book have been mastered, you will begin to intuitively know whether your cob is good, the wall is stable, the lime is the right consistency, and so on. And beyond this, the true creative process can begin. There is no better medium than cob with which to shape and literally sculpt a unique, beautiful, personal piece of art, and enjoy yourself at the same time.

Building with cob fosters a rejection of global homogenisation, monoculture, mass manufacturing, top-down solutions, and high-tech approaches. This age-old technique can be used in a truly fresh way to encourage regionality, the local, the specific, the appropriate, the low-tech, the simple. It can be connected to a larger movement that is going on in the world right now, that embraces home-grown and local production – whether it be indigenous music, local, organic and seasonal foods, or the resurgence of younger generations wanting to revive their local dialect and languages.

In this book we encourage you to go out and build something yourself that is highly relevant to the place and space that you inhabit, to become "tuned to the structure and pulse" (*The Spell of the Sensuous*, David Abram, 1997) of your particular place. This means responding to your environment, resources and needs, and building accordingly. For example, if you live in a wet climate like in the UK and Ireland, your house must be made from materials to withstand the rain and wind, and the roof must be suitably pitched to shed water from the building.

If you live somewhere where there is no clay subsoil, and where temperatures are excessively cold for long periods of time, then maybe cob is not for you (to find out why, read on). It is not a panacea for all buildings, in all places, for all people, and it should never be treated as such.

Consider also how, and by whom, your cob building should be built. In the past, it was almost always built by the people who were going to live in it – the ancient version of the if you utilise the skills of a professional cob builder, it can be an expensive option when compared with building a house made out of concrete blocks. The way we see it, there are three options for building something out of cob that are relevant to the structure of today's society:

> " A story that makes sense is one that stirs the senses from their slumber, one that opens the eyes and ears to their real surroundings, tuning the tongue to the actual tastes in the air and sending chills of recognition along the surface of the skin. To make sense is to release the body from the constraints imposed by outworn ways of speaking, and hence to renew and rejuvenate one's felt awareness of the world. It is to make the senses wake up to where they are. "
>
> *The Spell of the Sensuous*, David Abram

'owner-builder' concept. Today, society is structured very differently, and there are more choices around how such a building can be created.

Bear in mind that although the materials to make cob can be extremely cheap, if not free, it is a highly labour-intensive building process, and labour costs, if brought in from the outside, are expensive. For this reason, like most traditional crafts practised today,

1 Having a professional cob builder build your house for you – as discussed above. This can be expensive, but you are guaranteed a unique, functional and beautifully crafted structure.

2 Build it yourself. Cob is a perfect self-build material, being easy to learn and needing very few tools (no expensive or sophisticated ones). It is a great way to create community buildings through the

hands and feet of many people of all ages and abilities. Cob building is completely safe, and can be done by anyone who is willing to learn. Although needing someone with lots of time on their hands, by using local materials and your own labour with the help of friends, an extremely cost-effective building can be produced, which will foster empowerment and a great sense of ownership. Attending courses and reading this book should provide you with the skills necessary to build a cob structure.

3 Consulting with an expert. This approach falls somewhere between options one and two above, and involves hiring the services of a professional cob builder at the beginning of your project, to get you on the right track. This person can also consult at intervals throughout the building process, or as needed. This can reduce costs, because you can provide most of your own (cheap) labour, but can save your time and money by paying a fee to someone else who is an expert in the field, who can ensure that your project is destined for success.

This is how we see the way forward for building with cob in the 21st century. And indeed, as you read through this book, we hope that you will begin to appreciate its absolute relevancy for the modern world that we live in. As a modern, 'green' building material it has a head start on many of the other 'green' building technologies. It has been used all over the world for millennia, and many cob buildings – up to 500 years old – are still standing, and providing comfortable homes for millions of people. We know it works.

Here in the UK, we have our own very specific heritage of cob buildings, and can use these not just to imitate, but to learn lessons from and to find ways to improve on those things that need improving. Contrary to popular opinion, cob and other specific natural building techniques can meet the latest UK building regulations, which are there to encourage you to build an energy-efficient and safe structure, both of which cob can provide.

To come back to the beginning is to talk not just about circles, but more specifically, about the spiral. For it is not really the beginning that we want to come back to, but a different point on the same cyclical journey. For a spiral always returns to itself, but never at exactly the same place. Spirals never repeat themselves; they remind us that life is movement, and that nothing is ever the same. Repetition is neither sought nor valued, which certainly applies to cob.

Straight lines, on the other hand, measure; they are static, and they separate and divide. Cob is more dynamic. Cob is transforming, flexible, forgiving, empowering, practical, democratic, simple, inherently linked to the natural world, accessible, sustainable, renewable, beautiful and highly relevant to these interesting times in which we live.

Earth building around the world

1

Vernacular buildings record lifestyles of the past, when people had to find a sustainable way of life or perish. Just as we will have to now. The new importance of vernacular building is that it has vital ecological lessons for today.

David Pearson, *Earth & Spirit*

vernacular traditions and natural building

Mud has been used to create dwellings and structures since human beings first created shelter 10,000 years ago. It can be found in the simple shelters made of woven sticks covered in clay, the remains of which were discovered on the Nile Delta in Africa from 5,000 BC, to the rammed earth sections of the great wall of China, the majestic mud brick mosques of Djenne and Mopti in Mali, and the humble cob cottages of the British Isles. And before this, humans must have watched and learned from the swallows who weave their nests out of twigs held together by mud, and the termites who create huge mounds out of particles of earth piled delicately on top of each other.

The people making these buildings were (and in some societies continue to be) the children, women and men of the rural communities around the world. They were also the finest craftspeople of the world's most ancient civilisations, as well as the peasant tenant farmers of pre-industrial

Opposite: Earthen adobe Pueblo church in Taos, New Mexico. Below: Making mud bricks by hand in Merv, Turkmenistan.

Europe. Mud has always been, and continues to be, the most available, democratic and adaptive building material on the planet.

Vernacular building practices around the world

"Quietly and almost without notice, they outwit the might of modern machinery with simple tools and materials that welcome, encourage, and amplify the use of the human hand." Bill and Athena Steen and Eiko Komatsu, *Built By Hand: Vernacular Buildings Around the World.*

They are designed and built by the people who live in them, using the natural resources available locally, and using simple hand tools and a low-tech approach. They are designed to respond intimately to the local site on which they are built, and serve as an expression of the community's and the individual's cultural and social human needs. As Hughes and North said in 1908, regarding the vernacular buildings of Wales: "Just as the many-branched Welsh oaks are peculiar to the principality, so are these buildings the natural product of the country, the true growth as it were of the soil, and show as clearly as any

Earth dwellings, Africa.

30% of the world's population live in homes built of earth

50% of the population of 'developing' countries live in earth buildings

Earth has predominantly been used for building by the indigenous peoples of the world, who live in pre-industrial societies, who work and live off of the land, and have little or no access to our so-called 'modern' technologies. Vernacular building techniques are used for the homes of ordinary people.

written history the development of the life of the people." – Eurwyn William, *Home-made Homes: Dwellings of the rural poor in Wales.*

Vernacular buildings can be thought of as the equivalent to folk speech, local dialects, folk art and folk music – they are unique, specific, and their beauty lies in their simplicity, functionality, humility, and the fact that they respond intricately to the world in which people live. Much of modern housing – often necessarily erected hastily, as a response to the need to house an ever-increasing population – is lacking in this sensitivity. Often it would seem that modern developments are the product of visions created by designers and architects, who act on theories about how they perceive people should live. This can be seen in the tenement high-rises that were erected in the 1960s and 70s. They were born out of a social housing theory which, as everyone can now see from the ghettoes of the inner cities, was horribly wide of the mark.

Vernacular buildings are literally made by hand. Their beauty lies in their imperfections, irregularities, specific nuances and idiosyncrasies. It is ironic that most vernacular buildings, in which we find so much beauty, have often been made by people who have little money, no specialist knowledge, and who are simply striving to create shelter and protection from the elements with what materials they have got.

Most of the features that we find so desirable and beautiful in vernacular buildings, and that we strive to imitate in many modern 'designed' buildings, were born out of practicality and inherent common sense. Vernacular buildings are generally extremely efficient, and no feature emerges that does not serve a function – there was no room for embellishment.

We respond so deeply and positively to these features because we can feel and see the understanding that their creators had of the materials used and the environment in which they were building. It is ironic also that we now look to these buildings as models of 'green' practice in everything from siting, design, materials and methods. We have come full circle, and can begin to re-learn all that we have forgotten.

Earthen vernacular building in the UK, and the effects of the industrial revolution

"Probably indeed there is no county in the (United) Kingdom that has not considerable areas where the soil would, if tried, prove well adapted to cob building." Clough Williams-Ellis, *Cottage Building in Cob, Pise, Chalk and Clay*.

The simple labourer's cottage could be said to be Britain's indigenous, vernacular building. It was always built with materials specific to the region, but was predominantly made out of stone and mud from the fields to make up the foundations and walls. Local trees were used for the roof timbers, and the grasses and reeds from the surrounding area for the thatch roof. It was generally built by its owner with the help of the pooled labour resources of the community, which comprised the poor, rural workforce that served the local estate, owned by the landed gentry. These made up the homes of the ordinary people in pre-industrial Britain.

The onset of the Industrial Revolution in the 1800s brought dramatic changes to all aspects

(continued on page 18)

Top left: Ancient mud fortifications in Bam, Iran.

Top right: Prehistoric cliff dwellings, Arizona.

Above centre: Store compounds, Ghana, Africa.

Left: Minaret of mosque in Timbuctu, Mali.

Right: Earth dwelling, India.

Below left: Village building, Peru.

Bottom right: Cob house under construction, Oregon, USA.

World Heritage of Earth Building –
Past to present

Middle Stone Age
- 8000 BC: Oldest remains found so far, of earth structure at Jericho – town made of mud bricks.
- 7000 BC: India, on the banks of the river Indus – dwellings built with unbaked bricks. Fortress walls encircling village made of mud brick.
- 6000 BC: Oldest settlements in Europe – primitive dwellings on Aegean Coast of woven wood daubed with clay, and later sun-dried mud bricks.
- 5000 BC: Hunter-gatherer semi-permanent huts made of earth.
- first archaeologically recorded human settlements made out of earth on Nile Delta, Africa – woven reeds and branches covered with clay.
- China, dwellings dug out of loess clay in circular or oval shape.

New Stone Age
- 3000 BC: Athens – city established at base of Acropolis – earth brick buildings with thatched roofs.
- 2000 BC–2000 AD: China – earth brick as infill for timber frame.

Bronze Age
- 1800–570 BC: Mainland central Europe – wood and earth structures.
- 1600 BC: Peru Andean Region – earth structures and earth bricks for temple.
- 1552–1070 BC: Homes of artisans and nobles, palaces and temples built out of mud bricks in the New Kingdom, Middle Egypt.
- 1200 BC: Egyptian builders develop concept of mud brick vaulting – Lower Nubia.

Iron Age
- 650 BC: Central America – pyramid and houses built in earth.

Modern Era
- 300–400 AD: Rome – houses made of mud brick.
- 300–800 AD: Casa Grande, Arizona, USA – Native Americans in South-west of America – Hohokam tribe lived in houses of cob.
- 1100 AD: Islamic mosques constructed out of mud bricks, direct shaping or daub, depending on locality, e.g. Djenne and Mopti, mosques of Mali (both still remaining today).
- 1100–1300 AD: New Mexico, Arizona, North America – Anasazi Indians used widespread adobe brick construction.
- 1270 AD: Chapel La Salle de Diana – the oldest remaining complete earth structure in Europe, in Montbrisson, France. Now it is the town library for moisture-sensitive books.
- 1700–1800 AD: Earth building popular in Central Europe – Denmark, United Kingdom, Germany.
- Post-First World War: Earth building popular in Europe.
- Post-Second World War: Decline in popularity of earth building in Europe.
- 1972: UNESCO World Heritage Convention – three heritage sites included that used earth as predominant building material.
- 1974: CRATerre – Grenoble, France. Grenoble School of Architecture to undertake task of updating scientific and technical knowledge of unbaked earth construction.
- 1982: Exhibition & conference at Pompidou Centre, Paris, France, entitled *A forgotten building practice for the future*, about earth as a building material.
- 1993: United Kingdom, The Devon Earth Building Association (DEBA) formed.
- 1993–2006: Revival in Europe and North America of earth as environment-friendly building material. Earth remains the predominant building material in South and Central America, China and Africa.

> **Their pictures tell the story of a disappearing world of buildings that have been constructed by ordinary people who as builders and homesteaders have given artistic, modest, and sensible form to their daily needs and dreams. Sometimes accidental, often asymmetrical, and utilising materials that are naturally close at hand, these buildings, with their moulded curves and softened lines, convey a personal and human beauty.**

Athena and Bill Steene and Eiko Komatsu, *Built by Hand: Vernacular Buildings around the World*

of life. These directly affected how, where, and with what materials the homes of the 'ordinary' people were created. It brought about the beginning of the decline of vernacular building practices, and the onset of mass-produced housing. The Industrial Revolution created new factories in cities. They produced the standardised, machine-made and pre-fabricated materials that were used for all aspects of life, including buildings.

These factories provided a draw for much of the rural workforce; people moved from the country to the city in search of perceived improved living conditions and a hope for financial gains. The new factory workforce was provided with housing by the industrialists, which was pre-built using the new industrial materials such as brick, steel, and cement. These mass-produced houses consisted of back-to-back, identical structures that lacked the regional idiosyncrasies and individuality of the vernacular buildings. Artificial communities were created, and the rhythms of nature disrupted. Here began the decline of the owner-built home, and the ability of people to provide for themselves in all aspects of life. A new generation of specialists rapidly emerged, who began to lose touch with the well-rounded skills and practices of the generation before them. A new model of progress began to consume people's lives. Consequently, earth building in pre-industrial Britain began to fall out of favour. People began to have higher economic expectations, and the concept of modern architecture was born. A new set of social rules was established which gave way to style and fashion over pragmatism, and appearance over practicality.

This picture: Mud brick mosque of Djenne, Mali, Africa.

Others: Mud brick and earthen plastered mosques, Timbuctu, Mali, Africa.

Earth as a building material began to be considered as inferior, and the product of poverty. It fell out of fashion due to social reorganisation – not because it was less durable than the new modern materials.

Earth building as a solution to providing sustainable and environmentally sensitive building methods around the world

We are now more than 100 years on from the Industrial Revolution, and as a society find ourselves far from the practice of the vernacular building tradition of the pre-industrial era. The housing industry, with its highly processed, modern materials, now contributes to around 50% of all pollution in the world, and cement processing alone creates 8% of total greenhouse gases. The structure of society too has changed dramatically, and the once common practice of building your own home has all but disappeared amongst the majority of people living in the 'industrialised' world.

We live in a world of consumer abundance, and our inclination to be resourceful like the people who lived and continue to live off the land, has been temporarily eroded from our psyche. The changes that began to take place with the Industrial Revolution gained such rapid momentum and lulled us into such a false sense of 'progress' that we forgot to notice how deeply out of balance it was causing our planet to become.

The majority of processed building materials produce huge amounts of pollution at all stages of their production and life. Precious energy such as fossil fuels is consumed in vast quantities during their extraction, manufacturing, transportation and disposal. Their effects on the health of those who produce, install and live with these materials is being felt, as well as on the beautiful living creatures who have to deal with the toxic wastes that flood into the waterways, and seep into the earth.

A new generation is now emerging, of people who are engaged in a global search for alternatives and solutions to the state we find ourselves in; and these solutions are not proving hard to find. Some of these solutions can be seen in the buildings of the past, the structures of the still-existing rural tribes and communities around the world. They are in the very ground beneath our feet, and the grasses blowing in the wind, the sun that warms us, and the hands and feet that we are born with. Sometimes, the simplest solutions can be the hardest to fathom.

Earth building, along with other natural building techniques, is once again being noticed and valued as a practical and life-enhancing solution to the state we find ourselves in. Clay is a healer on all levels. It can heal physical trauma as a receiver of toxins, and can address all levels of society – the academic can analyse it, the scientist and engineer can test it, the poet can lyricise about it, and the child, woman and man can hold it in their hands and build their own home together, to suit their needs, to enjoy for a lifetime.

Imagine a building material that can be dug from or near the site; needs only the addition of locally grown straw, locally sourced aggregate and water; can be mixed with your feet and built with your hands. And when the building is no longer needed, it can fall to the ground, ready to be re-used by the next generation of natural builders. This is cob.

The case for natural building and a whole systems approach

If we look to our past to inform our future, we can begin to appreciate and be inspired by the values and practices of our vernacular inheritance. Our pre-industrial ancestors were

comes to building. But nature and people are dynamic, and we accept that we live in a very different world from how it was 150 years ago. Society is structured very differently, and it is impossible to deny the impact that the luxury of choice has on the decisions that we make. Thus, we do not advocate the return to an idealised and romanticised past, but we can adapt and improve the best practices that can help enhance our quality of life and lessen the load on our ailing planet. To simply try to rehash the past would be to miss the point altogether. Christopher Alexander, architect and author, talks about what makes a building alive or dead, and the conclusion that he

> We are seeking, learning, dreaming of the simplest and most beautiful form of human shelter made by human hands from the mother earth.

Nader Khalili, architect and earth builder

models of efficiency and resourcefulness, and in a time of dwindling resources and planetary poisoning we can begin to study and utilise many of their practices, especially when it

draws is that the buildings that are alive and feel good are those that ". . . are adapted, deeply, to land and to people . . . for it is not style that makes a building living or dead, but the freshness of its response to its surroundings." [Christopher Alexander, *The Nature of Order*]. We are therefore far from suggesting that imitating the cob cottages of the 1800s is the only desirable way to build. We are proposing a whole systems approach to building, which can be found within the principles of natural building, a movement that is gaining in popularity around the world.

Natural building involves not just what materials you build with, but how, where, and why. As natural builders, we are interested in both environmental and social sustainability. Central to the movement are many of the defining characteristics that we have discussed in regard to vernacular building practices, only this time we have made the decision to choose these elements, whereas our forebears had no other options. These elements include:

- An emphasis on the minimisation of the environmental impact of the building materials, practices and the building itself.
- A simple, low-tech approach wherever possible.
- The nurturing of a broad range of skills instead of specialist knowledge in just one area of building.
- The use of as many locally available and renewable resources as possible.
- A respect of the local environment on which the building is sited, and a unique and regional design that corresponds to this.
- The encouragement of the owner-built house.
- The use of predominantly natural

building materials, i.e. those that have not been industrially processed, such as stone, mud, straw, and wood.

Out of these will emerge a natural house that is comfortable to live in, healthy, beautiful, and life-enhancing – more than just shelter. Out of the natural building and whole systems approach come benefits for the planet and for people. A reduction in the contribution to pollution is high on the list of positives, and so also are the benefits it can have on human health for the builders and for those living in the buildings.

With natural materials it is possible to avoid the chemicals that are now commonplace in most building products, such as formaldehyde, glues and fibreglass products that can create cancers and chronic respiratory disorders. Also, the positive psychological impact of living in a natural home is derived from our response to the textures, shapes, irregularities and beauty inherent in the materials such as cob, lime, straw bale, thatch and timber. Finally, participating in the creation of your own home and with your friends, neighbours and community, no matter at what level, whether you are cooking sumptuous feasts to sustain the workforce or stomping cob, is incredibly empowering. To have handled and known every piece of a building, and to understand how all these pieces fit together, is about getting back in touch with our ability to provide for our families and ourselves. In this way, cob and natural building can begin to subtly and powerfully enhance ourselves, society and hence the world.

The different techniques of earth building throughout the UK

The use of cob extends across the south-west of England, from Cornwall to Hampshire. The term includes variants such as:

- Clob: Berkshire
- Wichert: Buckinghamshire and Hampshire
- Clom: South Wales
- Mud: Ireland
- 'Clat and Clay': Lancashire
- Clay Dabbins: Solway Plain
- Shuttered Earth: Most common in Norfolk and Suffolk. Also known as 'clay and bool' in north-east Scotland.
- Pise/Rammed Earth: most common in the chalk belt of Hampshire (fashionable in the early 19th century for houses of the gentry, therefore not strictly 'vernacular')
- Clay Lump: East Anglia

A selection of earth buildings, new and old, throughout the UK.

In the UK there are an estimated half a million inhabited earth buildings surviving in a range of types of construction and materials.

A history of earth building in the UK

Roman Times

St. Albans, Hertfordshire (Verulamium) –
archaeological remains of 'clean yellow clay'
Roman walls formed between boards
on stone foundations.

Viking Times

York, Norwich and London – archaeological finds
of low walls of clay, packed between shuttering
of wattle.

1200 AD: Earliest cob buildings in the UK recorded
from archaeological excavations.

1300: Oldest remaining cob buildings still standing
in the UK. Mostly the houses of middle status
members of society. Cruck timber frames (feet of
timber frame extending all or part way down to
ground) with water reed thatch or local slate.
Thickness of walls roughly 750–850 mm.

1600–1700: Simple A-frame timber roofs, thatched
or slated, with cob walls.

1775: Thickness of cob walls decreases to 600 mm
(24") or less, and wall heights increase.

Mid-1700s: Design began to become a separate
element of the building process. Cob still in
common use for construction.

1784: Brick tax encouraged proliferation of rural cob
buildings by landlords and builders.

Early 1800s: Cob used in public buildings,
chapels, schools. Variations in construction
methods.

1850: Brick tax was abolished and therefore more
brick was used.

1860s: Railways, sea and canal transport
began to be used to import building materials.

Industrial Revolution

Local materials (cob and thatch) were replaced
with manufactured brick, stone, fired clay tiles
and slate from Wales, especially in the urban
areas of UK. Brick production increased, due to
the mechanisation of the process. Standardised
industrial products began to be preferred over
vernacular materials.

1875: Arable farming decreased because of the
cheap grain available from North America, and
livestock farming also decreased because of the
cheap imports. Consequently, many farmers and
rural labourers moved to the city to work in the
factories, lured by the promise of a better life.
Building of farm cottage buildings (cob and
thatch) decreased greatly.

1911: Large house called *Coxen* built in Budleigh
Salterton, Devon, out of cob by Earnest Gimson,
the Arts and Crafts Designer. This house was built
by a cob builder, who had learned his cob
skills 30 years earlier.

1919: The architect Clough Williams-Ellis published a landmark book describing all earth building methods in the UK. He proposed earth as a solution to solve the shortage of building materials in post-World War I UK.

Post-World War II: John and Elizabeth Eastwick Field revised and re-published Williams-Ellis's book. Again this did not have a great effect on the building industry.

1978: Alfred Howard built a cob bus shelter at Down St Mary, initiating the revival of cob building in Devon.

1978–1995: Start of English Heritage's sensitive restoration of Bowhill in Exeter – a cob manor house originally built in 1500. This generated a great revival of building with cob.

1984: East Anglian Earth Building Association (EARTHA) founded to explore repair techniques for clay buildings in East Anglia, and to hold public demonstrations of these techniques.

1990: Alfred Howard builds new extension in cob.

1991: Devon Earth Building working group formed in response to the need for informed repair and maintenance of old cob buildings.

Plymouth University School of Architecture set up Centre for Earthen Architecture, to conduct studies and research into cob.

1993: Kevin McCabe builds new house out of cob – first entirely new cob building erected since the 1930s.

1993: DEBA (Devon Earth Building Association) emerges from the Plymouth University programme. They hold forums and print pamphlets.

1994: *Out of Earth* conference set up by Plymouth University Centre for Earthen Architecture, to search out opportunities of using earth as a contemporary building material. Experts from British Earth Building network and experts from around the world met to share experiences of earth as a building material.

2000: *Out of Earth II* conference held to address the need to develop better conservation techniques to ensure survival of earthen inheritance.

2006 and into the future: Cob is again being recognised as a viable ecological, sustainable and aesthetically pleasing option for a wide range of building applications – schools, community centres, homes, etc.

Site & design

Nader Khalili, *Ceramic Houses and Earth Architecture*

2

Chapter 2

There was a time when people created shelter for themselves that was always completely in harmony with the local, natural environment. All documented and existing vernacular and folk building practices around the world demonstrate the evidence of an intimate relationship with the outside environment and a deep understanding of the available natural resources. Shelter was (and still is, in many parts of the world) created from the materials found in the immediate area – wood, mud, grass and stone – literally raised from the ground into buildings of elegant simplicity. Shelter was always intelligently designed and sited to minimise the negative elements of nature such as extreme weather conditions and harmful predators, and to maximise the positive elements of the environment such as the warmth and energy of the sun to provide heat and light.

Modern building practices, in comparison, rarely seem to utilise, understand, or even consider the limits and gifts of nature. Many modern houses built by mass developers consist of soulless, square boxes, randomly oriented, and erected hastily out of mass-produced materials that have often been

Below middle: Plans for 'Cobtun' by Associated Architects. Building designed to face south for maximum solar gain.
Below right: Re-built cob carthouse, Cornwall, designed by Mathew Robinson, cob walls by Cob in Cornwall, thatch roof by Mike Pawluk, timber frame by Stefan Roux, stonework by Paul Finbow.

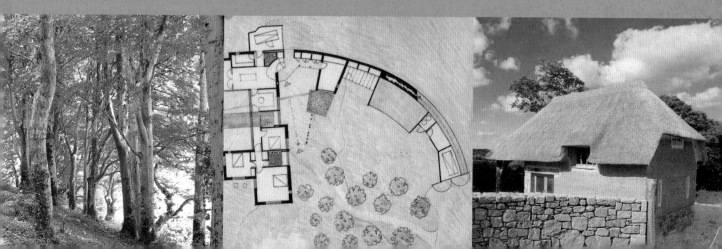

shipped from the other side of the world. Luxury features include 'sun-porches' facing north, randomly placed plastic bay windows with wonderful views of the neighbour's living room, roofing slates that have been shipped from China, and stud walls made out of timber from the dwindling rainforests of Brazil. One can build on land reclaimed from the sea, and create lush mansions in the middle of the arid desert. We seem to believe that we are no longer constrained by the limits of nature; and we no longer notice where the sun rises and sets every day, which way the trees sway in the wind, and what we can do with the earth beneath our feet.

We encourage anyone who is about to embark on creating something out of cob from scratch – be it a house or a modest garden shed – to re-engage with nature, and to take the opportunity to create something that really maximises the potential of the natural site: a building that seeks not to dominate, but to grow organically out of the environment in which it is cradled.

This building could be said to be truly adapted to its environment in the sense that Christopher Alexander describes. It responds to its surroundings and to the needs of the people who built and use it. The Myrtle, at the North American School of Natural Building, by the Cob Cottage Company, Oregon, U.S.A.

The Site

In the UK, it is a luxury to imagine being able to pick and choose the perfect site that neatly fits into all the ideal categories discussed below. We are constrained by two major factors: the sheer lack of land available on which one can get planning permission to build, and the outrageous prices that good building plots currently command. However, most building plots can be intelligently worked with to create a healthy, efficient, ecological and beautiful dwelling space.

There are some definite circumstances that one really must consider when siting and designing a natural building, which by definition attempts to minimise its dependence on artificially altering the land or relying on excessive use of artificial forms of heating the building. These include:

Good drainage

Never build on land that is marshy or situated within a flood plain, even if it is only liable to flood every 100 years. You may visit a site when the ground is dry, but ask questions about what happens during periods of wet weather. Talk to neighbours or the local authority about any streams that may be liable to burst. The ground should be well draining, with a solid, homogeneous sub-surface geology. If the land on which you want to build is very clayey, this is a bonus for making cob, but does mean you will need to pay good attention to ensuring that your drainage system is set up well to move water away from the building as quickly as possible, as clay is very slow draining.

Sub-surface geology

The sub-surface geology of the site on which you wish to build should be solid, and of one homogeneous material. Again, if your subsoil is very clayey, especially if it is composed of unstable expansive clays, you will need to dig past the clay layer and create your foundation on the solid bedrock, as clay will contract and expand when it comes into contact with moisture. If there are different substrates under the building they may move at different rates, which may create conflicting movement in the building, causing cracks and instability. Your building is only as good as the foundations it rests upon, and the foundation is only as good as the ground on which it sits.

Location

Avoid building on an exposed hill or coastal cliff, which may have stunning views but will come with the price of whipping gales and eroding salt air. Also, anything 500 feet above sea level will most likely experience severe weather in the winter, as this height is known as the theoretical snow-line. Anything built above this line will probably make access difficult in the winter, and will be more difficult to keep warm.

It is wise not to build in a valley floor, because your interaction with the sun will be limited, and in winter you will sit in a frost pocket because cold air travels downhill and comes to a standstill when it hits an obstruction such as a building.

Access

Consider the difficulties that may be involved with choosing a site that has limited or difficult road access. This will make transporting outside materials onto the site challenging, and it may become tiresome carrying groceries up a steep track on cold or wet winter nights. It is also absolutely necessary to ensure that a potential site is not landlocked – that it can be accessed without having to cross land which is owned by someone else.

Aspect

It is best not to build on north-facing slopes, or anywhere that has limited access to the sun. Psychologically it will not feel good spending most of the day in the shade, and it will be impossible to tap into the heat and energy gains from the sun which are essential to any dwelling that is seeking to limit its dependence on artificial forms of heating, fuelled by dwindling fossil fuels.

If the site is surrounded by trees which block out the sunlight, check whether they can be felled and utilised to build your house and heat your home, or whether they are of protected varieties. If you visit a site in winter, be aware that the leaves will not be on the trees, and may mislead you into thinking that the solar access will be good throughout the year.

Legal restrictions

Before you purchase or build on any piece of land, ensure that there are no legal restrictions that could prevent you from creating exactly what you want on that particular site.

History of the land

Ensure that the land on which you wish to build has not been contaminated by previous owners and their activities, which may have involved the disposal of noxious or toxic waste on or near the ground. It is a good idea to get a contamination report done, which may be required anyway by the local authority when seeking planning approval. Learn as much about the history of the land as you can from neighbours and local records. However, building on a piece of land that has been scarred or misused in the past, as long as it is not dangerously contaminated, can be a very positive thing to do. This can be an opportunity to repair a damaged and depleted area, which has been an eyesore, and turn it into a place of beauty, where wildlife can flourish.

Future developments

Before purchasing land, ensure that you become aware of any future developments for the surrounding land that may directly affect your quality of life. These may involve new roads alongside your back garden, a bypass, or a housing or industrial estate nearby. Find out the local authority's plans for any adjoining land. This will prevent you from waking up one morning to discover that the adjacent horse pasture is being turned

into a smelting factory! Useful websites that may help you to answer some of the above questions are:
www.environment-agencies.gov.uk
www.upmystreet.co.uk

There are some other elements to consider when choosing a site to build a house or a small structure, which will greatly enhance both the efficiency and functioning of the building and the psychological experience of living in and around the space.

Most of these are pure common sense, and have been practised out of necessity since human beings began to erect shelters. Most of us have lost touch with these innate sensibilities because we are now able to rely heavily on artificial forms of heat and lighting for our houses, and have access to whatever building materials we desire, whether it be York Stone in Surrey or Welsh slate in Cornwall. Traditionally, though, shelters relied on good siting as a means of survival, and to make the difference between a miserable life in the cold and wet, or one of comfort and refuge in the warmth and dry.

Looking to nature

One only need look a little deeper into the habitats and dwellings of the small creatures in our back gardens to notice the intelligence behind where and how they site their homes. Consider the thirteen-foot termite mounds of Australia, which are oriented on a precise north/south compass line. The termites spend time on the east side in the mornings to capture the warmth and light of the rising sun, and then move to the west side in the afternoon, as the sun moves around and sets. Also, consider the beehive, which consists of hexagonal tubes making up the honeycomb. This shape encloses a large amount of volume allowing for maximum module space and a minimum of surface area exposure to the cold. (These examples are taken from *A Shelter Sketchbook* by John S. Taylor). This may seem like awe-inspiring poetry to us, but simply illustrates the survival instincts of all creatures living in the natural world.

Orienting towards the sun

How can we re-learn these skills? The first place to start from is to imagine building a shelter without any access to mains water, electricity or gas hook-ups. Then consider what your main priorities would be to lead a comfortable and fulfilling existence within the space. In a British climate, most people would consider being warm and dry, especially in the winter months, as being high on the list of priorities. With this in mind, we can quickly come to the conclusion that the sun is going to be the most important organiser of the building orientation. As touched on briefly before, this entails ensuring that the site, and hence the building placed on it, is oriented so that it faces as much as possible between solar south and south-west, and therefore has unobstructed access to the day's sun between 11 am and 3 pm to heat the inside of the building. This follows the ancient principles of passive solar design, and will be discussed in more depth later in the chapter.

Utilising the natural resources on the site

Just as important to consider is the question of what sort of natural resources are available on the site from which to create the structure. It is no coincidence that the existing materials are going to be those most suited to the climate of the local area, ensuring that the structure can withstand the elements and provide the most comfortable space within which to live for that particular place. For example, the ancient adobe tradition that exists in the hot desert climate of the south-west of America consists of sun-dried earth bricks which have the ability to absorb heat and hence balance the hot day temperatures with the freezing night-time temperatures. The soft mud tones of the walls and roofs echo the surrounding desert, so that they become a part of their landscape, like a termite mound or a rocky outcrop. Is it a coincidence that the lack of trees in the desert, which are needed to make the frames for the pitched roofs used in rainy climates such as the UK, coincides with the fact that it rarely rains in the desert, and roofs can therefore withstand being constructed out of the same earth bricks as the walls? Consider also the pine timber cabins of Scandinavia, which are kept deliciously warm by the wood's natural ability to insulate well against the harsh winters. The steep-pitched roofs, designed to effectively shed snow, mimic the shape of the pine tree forests that they are surrounded by, so that these buildings again seem to grow out of their environment, like the trees they are made of. For the owner-builder of a cob structure, the realistic utilisation of on-site resources may include good sources of building stone for the foundations, trees that can be sensitively felled for roofing structures, a good clay subsoil with which to make cob for the walls, and nearby fields of wheat-straw to make a thatch roof. If there are resources on your site that can be utilised, you also need to consider where on the site these resources are in relation to the potential site of the building, and hence the ease with which they can be potentially transported. Bear in mind that it is much easier to move heavy materials such as barrows full of cob or heavy stones downhill to a site, rather than uphill.

Nowadays, unfortunately, the idea of being able to extract from the land all the resources needed to build a complete structure or dwelling that complies with current British building regulations, is simply not a feasible one. And indeed, one must always remain fully aware of the ecological impact you would have on the site by removing all the trees to make a roof, or by upturning every available stone to create a foundation. It is important to respect the occupants who already inhabit the land – from the beetles which live under the rocks, to the ancient oak tree that may stand in the ideal spot for a house.

Peaceful observation and intuition

After taking all of the above practical considerations into account, remember that it is up to you to feel out and to discover where your building would best sit on the land.

The steep-pitched turf roof of this Icelandic cabin mimics the shape of the mountain backdrop.

Cob allows you to create free-flow forms and unique designs. Cob courtyard by *Seven Generations Natural Builders*, Oregon, USA.

There is nothing more powerful or revealing than simply sitting quietly on the site at different times of the day – and if possible of the year – to observe the patterns of the sun and the wind and discovering where you are intuitively drawn to. You will probably find that the land draws you towards the very spot that contains all of the right criteria for the creation of an efficiently functioning, healthy and satisfying living space.

Building a structure on an undeveloped site provides a good opportunity to establish a reciprocal relationship with the land, with a consideration of the needs of the site as well as one's own needs. Seek out what the site has to offer, instead of solely imposing your ideas of what you want the site to do for you. During the actual construction process, try to keep damage of the surrounding area to a minimum, and try not to destroy beauty in your search for it. As Christopher Alexander says in his book *A Pattern Language*: "Leave those areas that are the most precious, beautiful, comfortable and healthy as they are, and build new structures in those parts of the site which are least pleasant now."

Designing with cob

Once you have picked a site and decided where to place your building on the site, the exciting prospect of designing the actual structure begins. Designing a building to be made out of cob really can liberate you from the conventional, square house designs that have become the norm of modern architecture.

Cob is a material that allows you to create shapes and forms that the more rigid, less malleable materials such as cement block and standard milled wood will simply not allow. This enables you to be truly individual and personal with your design, and gives you the opportunity to create a living space that really fits in with your lifestyle and your own unique needs and desires.

Creating plans with an architect

To create a dwelling or larger structure in the UK which requires planning permission and building regulation compliance, you will need to submit a full set of scaled drawings to the local authorities. They do not need to be produced by a professional architect, as on the

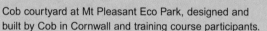
Cob courtyard at Mt Pleasant Eco Park, designed and built by Cob in Cornwall and training course participants.

This cob courtyard (same as left) is positioned on a blustery hill close to the sea. The thick cob walls provide shelter from the prevailing winds and allow for solar gain from the south. Scantle slate roof by Ben Verry.

continent, but need to be neatly and accurately drafted to illustrate even the smallest of details. The most important consideration when designing any building is to get to know the limits of the materials which you are using. It is therefore important, if you do work with an architect, to work with someone who has experience with cob – its limits and its endless possibilities – and with someone who is passionate and committed about creating buildings out of natural materials that will leave a minimum footprint on the planet.

There is a growing number of 'green' architects and designers who can fulfil these aims. Be warned of architects who design buildings solely to fulfil their own whims and indulgences! The best relationship to be had with an architect or designer is one where a trusting partnership can be fostered, where you can draw on their knowledge of buildings to translate your inspirations and ideas onto paper. This should produce a building that is self-inspired but structurally sound.

There are three main things to consider in the design of any natural building:
• How it looks from the outside, particularly

in its relationship to the larger context of the environment in which it is situated.
• How the inside space works in relation to your needs and desires.
• How it can be designed to (a) maximise the energy from the sun to heat the inside space and (b) protect the inside space from the cold in the winter, and overheating in the summer.

Building aesthetics and blending in

This is of course down to personal preference. However, when designing a natural building one is generally aiming to create a building that is sympathetic to, and responds to, its surroundings, so that it becomes a part of the landscape instead of being an imposition on it. Natural buildings aim to grow out of the spirit of a place. Refer back to the steep-pitched log cabins of Scandinavia and the adobe desert-houses of the USA, which mimic their surroundings because they are made primarily out of their surroundings. The traditional cob, stone and thatch cottage is a prime example of a naturally adapted building, because it nestles snugly into the green, brown and golden palette of the British countryside from which it is made.

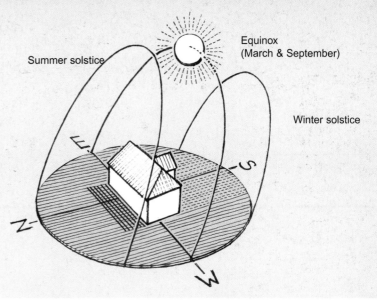

Seasonal fluctuations

The most effective passive solar building is rectangular in shape, one room deep and has its longest side oriented from East to West.

Summer solstice

Equinox (March & September)

Winter solstice

The sun as a heat and light source and organiser of space

The second and third points stem from the ancient, simple, but fundamental concept of passive solar design. Passive solar design has been used in various forms for millennia, and simply refers to designing a building so that it can effectively utilise the heat of the sun to create light and warmth inside the building. The basic principle relies on the transmission of the sun's (radiant) heat through a glazed layer directly into the building space. Some of this heat can be used immediately in the living space, and some can be stored in the thermal mass of the floors and walls and then re-released slowly when the temperature inside the building drops, such as at night time.

When designing a building so that it can utilise these passive solar principles, the following design elements must be taken into consideration early on in the planning:

1 *The orientation of the building, and the placement of the windows within the building* – their locations and direction to allow the sun's rays to penetrate to the inside space.

2 *The capacity of the materials from which the building is made,* to absorb the heat from the sun's rays and then re-release it into the inside living spaces. Materials to be used for this purpose must have good 'thermal mass' (see below).

3 *How well insulated the building is,* so that it is able to retain the heat gained from the sun, which has been absorbed by the high thermal mass materials. Without good insulation, the heat gain from the sun will be lost as quickly as it was gained.

Building orientation and window placement

The building orientation is discussed in the previous section on siting, and simply requires that the main living spaces of a building be situated so that they are facing between solar south and south-west. The windows of a passive solar house should also be placed predominantly on the walls facing between south-east and south-west.

To avoid overheating in the summer, they should not be placed too high up, to avoid overheating when the sun is high in the sky.

(continued on page 36)

Working with architects: an interview with an environmentally-informed architecture firm

ARCO$_2$ Architecture Ltd is an RIBA Chartered Practice, and is therefore governed by the RIBA and the ARB Codes of Conduct. Initially a client will make contact with our practice and we will either arrange to meet with them on site or at our eco-office conversion. The first meeting will involve discussions regarding issues such as the nature and scope of the project, the site, procurement method, budget, time-scales, feasibility, planning policy, other professionals and consultants required such as structural engineers and planning supervisors, etc.

Following the initial contact we provide the potential client with a detailed quotation letter, which also helps to formulate a basic brief. Architects have to be appointed in writing with either a standard form of agreement (SFA) published by the RIBA or via an appointment letter. The RIBA have a system that enables both architects and clients to track the development of a project from start to finish, known as the Plan of Work stages A-M. Often clients find the work stages to be complicated.

Once appointed, we usually ask that a client provide us with a general brief; the brief is perhaps one of the most obvious and useful tools required for an architect to design a building. The more detailed the brief is, the better informed the design will be. Depend-

ing upon the client and nature of the project, the brief can evolve as the project develops. Once we have identified the client's requirements, a further site visit is made to undertake detailed site analysis work. This work will often include orientation, siting, natural light, solar path, prevailing winds, shelter, privacy, gradient, views, vantage points, surroundings, as well as other criteria. This initial analysis work is crucial to good sustainable and conceptual design. Often a concept is generated whilst on site.

The next step is undertaken within the office, which is a sketch scheme. Depending upon the client's resources and type of project we often research planning policy, existing services within the site and any restrictions. Once this is underway the sketch scheme can begin. We use a variety of techniques to generate proposals including freehand conceptual sketches, 2D Computer Aided Design, 3-dimensional models, animations, physical models, perspectives and photomontages.

Some of this work is also dependent upon the clients' architectural background; in certain cases they may specifically ask for proposals to be presented in a certain medium. We aim to design new buildings, which are entirely site-specific. The building should belong to the site and the site to the building. Our designs are derived and informed from the site analysis work, which is often evident within the concept. However, this is not always the case, for example with existing buildings. Once the initial sketch scheme is complete, a presentation of the proposal to the client is arranged.

This period is an exciting time for a client; to see their ideas and dreams translated into a design for the first time. We often suggest that the client provides us with written feedback so that they have time to fully reflect upon the proposal. Following this stage there can be many amendments to the design until the client is completely satisfied, or none at all.

The next stage is to arrange a preliminary meeting with the local Planning Authority. This is almost always undertaken within the local council offices, and is only usually an officer's informal opinion. However, this phase is necessary to establish whether the council is likely to support the application, or if the proposal causes any controversy. This can be extremely useful to ascertain whether other specialists such as arboriculturalists or planning consultants are required.

Following the initial planning consultation, the proposal may require further design amendments that will be developed with the client and architect. We then prepare a Pre-Application package, which is submitted to the local planning authority. This process can take several weeks, but is invaluable because this is often an indicator as to whether the project is likely to gain planning approval. The planning officer will respond in writing stating any planning policies that may prejudice the future application.

Once a written response has been received, we prepare final designs with the client, and also produce supportive planning drawings and documentation as required. The application is then submitted; both client and architect then have to wait for the application to be determined by delegated powers within the council or by a committee decision.

This will also ensure that the low winter sunlight will penetrate into the building. The north wall or 'cold wall' (so-called because it never receives any direct sunlight) should have few, if any, windows placed in it, because they will simply provide exit points for heat to escape from the inside of the building. Some windows can be placed in the east to allow the welcoming morning sun to fill sleeping and breakfast spaces with cheery light and warmth. Windows in the west wall should typically be small and few in number. Large west-facing windows will cause overheating inside the building in the summer, especially when the sun begins its descent.

Glass is an extremely effective conductor of heat, and will therefore effectively capture the heat from even the coolest winter sun. Glass will allow sunlight to pass through it to produce light and warmth inside the building, and then will trap this heat by absorbing some of the infra-red radiation. This is the same principle as the 'greenhouse effect' currently taking place on the planet. However, normal glass will quickly relinquish this heat as soon as the outside air temperatures decrease, so it is essential that the glass be at least double-glazed, if not triple-glazed, so that the radiant heat from the sun can be prevented from being quickly lost. There are many glazing products on the market with excellent credentials which will vastly improve the effectiveness of this heat collection (see *Resources and Suppliers*). Large panes of good glass are more effective than small panes, because their ability to insulate

will surpass the insulative ability of the frame. The best window frame material is wood, because it has a much better insulative quality than PVC, aluminium or steel frames. To further improve the ability to keep this heat inside the building when the temperatures decrease, especially at night, thermal curtains, internal or external shuttering, or blinds should be used. Conversely, shades may be necessary in the summer to prevent overheating due to the increased amounts of incoming solar radiation.

Glass is an extremely energy-intensive product to create, but the energy savings it can produce via a passive solar system, through the creation of heat and light, will balance this.

In Britain, where we have a large number of cloudy days in the winter, and where the direct sun needed for passive solar systems can be in short supply, it is necessary to be able to capture the diffuse solar radiation through the clouds, which is emitted from the whole surface area of the sky. For this reason, it is beneficial to have some roof windows to capture this. To be effective, however, they must be openable for summer ventilation, have summer shades to prevent overheating of the inside space, and be made of well-insulated glass, so that inside heat does not escape in the winter.

With these points in mind, it makes sense to make a rectangular building (not square), that is one room deep. This will maximise the amount of sunlight that is able to penetrate deeply into all the rooms on the inside space of the building. The long side of the rectangle should be oriented from east to west. A study of the old cob cottages around where we live, reveals that most are indeed only one room deep, are rectangular, and are oriented with their long side from east to west.

To enjoy the light and warmth of the sun as it travels on its path throughout the day, it is possible to design your inner spaces in accordance with this. Breakfast rooms and bedrooms are best positioned facing the east so that they can be filled with light as the sun rises. Living rooms and conservatories should face south so that they receive maximum solar gain throughout the day, for heat storage in the thermal mass of the cob walls and floors. A sheltered west-facing nook will make for the perfect spot to sip tea in the late afternoon or evening, and to watch the sun make its descent.

Thermal mass materials for maximum heat storage and effective heat distribution

To effectively utilise the warmth from the sun's rays to heat the interior of the building it is essential that the materials with which the building is created, especially the walls and floors that directly receive the incoming solar radiation, have good thermal mass properties. Cob has excellent thermal mass properties, which means that it can absorb the heat of a sunny day both directly from the sun's rays, and indirectly from the warm air. This heat will be stored within the thermal mass and then slowly released into the inside space

once the temperatures drop. Thermal mass materials will also go some way to preventing the inside air temperature from overheating in hot temperatures, because the majority of the heat will be absorbed into the thermal mass and not into the internal space. Thermal mass materials are therefore great equalisers of temperature because they are able to harness the heat of the middle of the day and then release it overnight, thus preventing overheating in the building in the mid-day period. The thermal mass of an earthen floor will conduct heat directly into your feet (or your whole body, if you lie down!)

The heat relinquished from the thermal mass store is distributed inside the building via radiation from the heated surfaces, and by natural air convection currents through open doors or high-level vents between rooms.

It is advantageous to have as much thermal mass material within a building, including walls and floors. Earth and stone are two of the best thermal mass materials, so cob walls and earthen floors are ideal to be used within a passive solar system.

Maximum insulation

Despite having excellent thermal mass and therefore being able to store a lot of heat over long periods of time, earthen walls are not good at preventing the loss of heat. They are therefore not very good at insulating against cold outside temperatures (see Chapter 8: *Insulation* to find out why). For the execution of passive solar principles to be successful, it

Sunlight hitting the thermal mass of this fired-clay-tile floor is absorbed and re-radiated when temperatures drop, creating an even, comfortable internal environment for this spacious meditation hall in Chithurst.

is essential to prevent the excessive loss of the precious heat from the sun that has been captured inside the building and absorbed within the thermal mass of the cob. To compensate for the poor insulative values of the cob, it is essential to over-insulate other areas of the building – the roof, floors, foundations and stem wall. Heavy thermal curtains will help prevent heat loss through the glass of the windows in the winter, as will ensuring that all windows and doors are tightly fitted and sealed.

In most parts of Britain, where winter temperatures do not often stay below freezing throughout the day, the above solutions will be adequate to ensure that the building is insulated adequately to utilise the benefits of passive solar heating. In climates where temperatures are often below freezing, hybrid buildings, combining cob with straw bale walls in the north elevation, will provide a good solution (see Chapter 8: *Insulation* for details on this).

Designing passive solar systems with cob

Sunspaces, greenhouses and conservatories can all be incorporated into a cob building to utilise the best properties of cob as a heat store with the principles of passive solar design mentioned above. A sunspace simply consists of a large south-facing glazed section of the wall, which will receive solar gains throughout the day. These solar gains will be stored in the thermal mass of cob floors and walls and released gradually when temperatures decrease, such as at night-time.

A conservatory or greenhouse can be attached onto a cob building as an additional living space, and separated from the main building either by a solid cob wall or large, opening glass doors. Such a space would act as an effective solar collector and serve many other positive functions to enhance daily life, especially in the colder winter months. These include providing a dry, accessible area to store wood, an area to dry washed clothes, as a buffer against cold outdoor air temperatures, as a 'mud room' where muddy boots and wet clothes can be taken off and dried, and as a pleasant light and warm space to relax and work in, when it feels too cold to go outside.

As mentioned above, these glass-rooms need shading devices and openable windows at high and low levels, as otherwise they will overheat in the summer, especially if built against a solid cob wall. They must also be double or triple-glazed, with thermal curtains or shutters in the winter, to protect

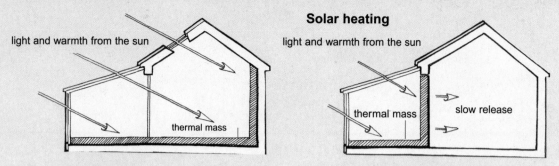

Solar heating

light and warmth from the sun

light and warmth from the sun

thermal mass

thermal mass

slow release

A south-facing glass conservatory attached to a cob house will maximise solar heat gain and storage in the mass of the cob floors and walls. This glass room can be separated from the building by glass or a solid cob wall.

from heat loss. Such systems work best when there is a rapid response heating system working off a thermostat, installed in the main solar-heated living room of the main building. This is essential so that the heating system can respond quickly to changes in temperature incurred by the potential solar gains of the day. A wood-burning stove or solid fuel boiler, for example, cannot be quickly turned off when the sun comes out on winter days. A heating system such as a thermostatically controlled wood-pellet burner is ideal.

Ancient wisdom in the modern world

Passive solar principles, or designing around the sun, are not new concepts. Examples can be seen in simple but effective details of all vernacular building styles. For example, the bevelled window openings of cob houses, with a wider angle on the interior walls of the building, were intelligently designed to enable more sunlight into the building without having to increase the size of the window opening, which would encourage heat loss. In the fourth century AD the Romans used large south-facing windows to heat their public baths.

Given the predictions about our future climate and global warming, it would also be wise to consider designing not only passive solar heating systems but also passive solar cooling systems into all new buildings. The thermal mass properties of cob as an effective heat store described above, will also work in the converse to prevent the overheating of a building. For this reason, cob is a material worth investing in for the future.

There are many sources for guidance and inspiration that can take you deeper into the practicalities and the lore of how best to site and design a new dwelling. One such example is the practice of permaculture, and the many practical and common-sense ideas relating to the intelligent siting of human dwellings that it offers up. These principles have at their core the idea that buildings and homes can be integrated harmoniously with plants, animals, soil, and water to produce stable and productive communities.

These can be studied and drawn upon to grasp a deeper understanding of the subject. But before modern interpretations of design and siting concepts, such as permaculture, were born, there were simply groups of people, in different parts of the world, simultaneously using their common sense and ingenuity to survive and flourish, utilising nature to provide warmth, comfort and shelter.

Identifying & testing soils

3

A subsoil consisting of roughly 15–25% clay and 75–85% aggregate makes for an ideal cob mix.

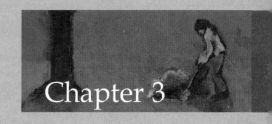

Chapter 3

Cob is made from the simple soil beneath our feet – but not the topsoil from your garden, which is teeming with life and organic matter. It is made from the subsoil layer underneath the topsoil, which may or may not contain in varying proportions the essential ingredient in cob: clay. To make a good cob mix, to produce a strong, stable building, your raw material should ideally consist of a ratio of roughly 15–25% clay and 75–85% sand/aggregate. The subsoil that you dig up may contain different variables of clay, sand/silt/aggregate, and hence you may need to add more sand/aggregate or clay to achieve the above ratio. The following is a range of tests that will help you to identify the composition of your subsoil, and show you how to amend it if required.

A brief breakdown of soils for making cob

All soils are made up of different proportions of gravel, sand, silt and clay, but not all soils have some of each grade. Scientifically, all soil types have been classified in relation to these proportions to describe their characteristics, such as a sandy clay, pure clay, sand, loamy sand, etc. The gravel, sand and silt element of a soil relates to the different-sized particles of rock present. They are the stable portion of the soil, as they remain the same size when wet and when dry. To be useful as a building material, they need a binder such as clay.

Below: The rich golden tones of a clay subsoil.

Clay is inherently unstable – it swells when it is wet and shrinks when it is dry. Each clay particle is surrounded by a film of water, and this is how the clay can bind together with its neighbouring particles and produce a solid mass. As clay dries out, these neighbouring particles are sucked very closely together, which is why clay shrinks when it is fully dry.

clay can be found locally: old quarries, a neighbour who is digging a pond, a farmer's field (farmers are good people to talk to as they work the land daily); and our favourite option is to utilise clay subsoil dug up to make way for the foundations of a housing development on a building site or road works. This 'waste product' of the building industry is carried off to a landfill and

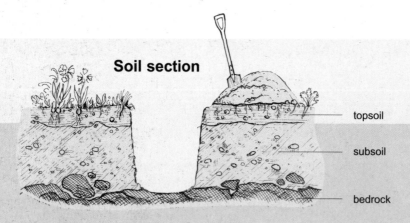

Soil section

topsoil

subsoil

bedrock

Soils change in their nature and composition as you dig deeper into the ground.

The soil needed for making cob will be found in the subsoil layer, underneath the topsoil.

Nature's clues

Clay subsoils are present in many areas of the UK. Nature gives us clues as to where we may find these clay deposits: look to areas where water comes to the surface, or where water sits for a long time after a rainstorm; also the areas in your garden which are notoriously hard to dig, and where the ground cracks when dry.

The ideal scenario is to source the subsoil from your own back garden or the plot on which you hope to build. If this is not possible, there are many other places where

dumped at a cost. They are generally more than happy to offload it to you for free, thus avoiding their dumping charges. Mention of a few beers or a bottle of whisky and they may even deliver straight to your site.

If you end up having to source the correct material out of the radius of your local area, consider whether cob is the right material for you. The essence of cob building has always been to source the materials as locally as possible, to produce a structure that is literally in and of the immediate environment.

Excite your senses

Utilise your senses of touch, smell, taste and sight. These simple tests may save you from doing any more elaborate tests if your soil is found to be unsuitable at this stage.

Visual

Clay normally comes within a certain palette of colours. These can range from a golden brown, oranges, deep reds and pinks, light pink, grey and mauve. There may even be pockets of different colours within the same deposit. Whatever colour the clay is, there will always be a distinct change from the visual colour and texture of the topsoil, which is normally a dark brown or black colour. You will also know you have hit clay subsoil when your shovel starts to leave shiny streak marks in the soil as you dig. Be careful here, as silty soils can also leave this shiny surface. A predominantly silty soil is not suitable for building with. Be sure to continue testing, using your sense of touch.

Touch

Take a small sample of your subsoil and crumble it between your fingers. Spit on it to create a damp paste. Clay will feel greasy and smooth to the touch, whereas silt will feel floury and crumbly. If there is clay present, it should feel tacky and make your fingers stick together. Once your hands are covered, wash the soil off using water – if it is difficult to clean off, it has a good clay content; if it washes off easily and doesn't feel sticky at all, then your soil is too sand-rich and does not contain enough clay to build with.

A clay-rich soil will hold its form well in your hands and will be easy to mould and sculpt into balls and shapes. Roll it into a sausage; if it is clay-rich it should wrap around your finger without breaking.

Taste

This is for those who really want to get intimate with their soil! Put some in your mouth and grind between your teeth (soil is normally clean), a smooth texture indicates a clay-rich soil, a gritty texture indicates a sand-rich soil.

Smell

As soon as you have freshly excavated your soil take a whiff – a mouldy or musty smell indicates the presence of organic matter, which suggests that you may still be in the topsoil layer, whereas a clay-rich subsoil smells clean and chalky.

Above: The kids at Padstow Primary school fully immersing themselves in the cob at a workshop to help build their new cob and straw-bale music room. Designed by ARCO$_2$ Architects, cob and lime work by Cob in Cornwall.

The shake test

1 Dig three or four holes for sampling, digging below topsoil to reach subsoil.

2 Place samples of subsoil in labelled jars ensuring no topsoil is present.

3 Fill jars one-third full with samples, and add one teaspoon of salt to each sample.

4 Add clean water to just below of the top of the jar.

5 Tighten lid and shake vigorously for 30 seconds.

6 Rest jar on level surface, allow sedimentation to occur, then evaluate.

Ensure that your soil sample does not contain any organic matter from the topsoil, as this will confuse your results.

You will know when you have gone past the topsoil layer and reached the subsoil by the change of colour from the dark brown/black of the organic matter to a range of potential colours – golden brown, oranges, pink, grey.

The subsoil should also be harder to dig than the topsoil.

Break up any clods of subsoil and remove large stones before placing in the jars. Adding salt to the samples will help the clay to settle out faster. A blob of liquid soap will also do the job. Make sure the lid is screwed very tightly onto your jar before

you start to shake vigorously! Whilst the sample is settling out, the jars must not be moved for at least 48 hours. Jiggling the jars about will invalidate the results.

Once everything has settled out and the water becomes clear, use a felt pen to mark out the different layers. This will help you to determine your ratios more easily.

Simple field tests

There is a wide range of tests that you can carry out to analyse your soil: from very simple sensory tests such as basic observations, to simple field tests, to the more complicated in-depth laboratory tests. It is possible to glean a good enough knowledge of your soil for building via the sensory tests coupled with the simple field tests.

If you lack the confidence to trust your own judgement, however, or you are really interested in getting a detailed breakdown for your soil, you can get a professional soil sample done in a laboratory at a local university. The Earthen Architecture Department of Plymouth University, as well as Bath University, can provide this service.

To obtain an accurate sample observe the following guidelines:

1 Always use subsoil, never topsoil.

2 Always test several samples from the same plot, as soils can vary greatly over small distances.

3 Always remove large stones from the sample.

4 Always break up any lumps present in the soil to create a consistent sample.

Historical clues

The first and most simple place to start is by looking at your local vernacular building history. Is there a tradition of earth building in your area? If not, why not? (no clay deposits?) Where are these buildings concentrated? Have they stood the test of time well? Most regions of the UK have a tradition of earth building. Talk to members of the older generations, and ask about any memories they may have about the earth buildings in your area.

Shake test (opposite)

This is the most revealing of the tests, as it will not just tell you if there is enough clay in your subsoil, but will give you some rough percentages of the overall composition of the soil. The final results of the particle sedimentation will help you to analyse how close your soil is to the ideal 15–25% clay and 75–85% sand/aggregate ratio, and therefore show you what, if any, amendments need to be made. Follow the six steps at the beginning of the chapter, place your jars down on a flat surface where they can sit undisturbed for a few days, and then immediately begin analysing your results as follows:

- Watch the samples in the jars immediately after setting down, and take note of how quickly each sample settles out into the different components of the soil: clay, silt, sand and small stones.

- **Within ten seconds:** small rocks and coarse sand will settle to the bottom of the jar.
- **Within ten minutes:** Fine sand and silt will settle out above the rocks and coarse sand.
- **Within a few hours to several days:** hopefully the liquid above these layers will remain brown and cloudy for some time, which indicates that there is a presence of clay. Leave the jars undisturbed until the clay begins to settle out.
- When the clay has finally settled out and the water turned clear, carefully make markings on the side of the jar (don't disturb the soil) at the junctions where each different element has settled to help give you a clearer picture of the composition of your soils.
- If the liquid at the top of the jar is clear immediately after shaking or within half an hour of setting the jar down, there is sadly no clay in your soil and you will need to source different clay to build with.

You can analyse the ratios of each component, and from here assess how much it departs from the 75–85% clay and 15–25% sand/aggregate parameter, and how you may need to amend the soil.

For example, if your soil sample shows a high ratio of clay, you will need to add extra sand/aggregate. Conversely, if your soil displays a high sand/aggregate ratio you will need to add more clay, or find another soil.

The shake test, along with the other tests mentioned, should give you a very comprehensive and sufficient picture of your subsoil.

Leg test

Take a handful of subsoil, dampen and make into a paste, rub some onto your leg and leave for a couple of hours. Try to rub it off – a silty or sandy soil will dust off easily with little effort, a clay-rich soil takes some intensive rubbing.

Sausage test (opposite)

1 Grab a handful of subsoil, moisten and roll into a sausage about the diameter of your thumb and about 150mm (6") long.

2 Place this sausage across the palm of your hand, and with your thumb gently nudge the end of the sausage over the edge of your hand.

3 Observe at what point in the sausage it breaks off, and measure the broken piece to give an indication of the composition of your soil:

- If the broken piece measures 130mm (five inches) or less, your soil has a low clay content.
- If it measures 130–250mm (five to ten inches), you have a medium clay content.
- If it measures 250mm (ten inches) or more, you have a high clay content.

Getting the mix right

water

clay (1 hour to several days)

silt and fine sand (10 minutes or less)

coarse sand & aggregate (10 seconds)

Mark the jar at the positions at which each element has settled, to interpret the percentages of sand/aggregates, silt and clay in the sample.

Right: To build successfully with earth you need the right amount of clay, aggregate, sand and silt: roughly 15–25% clay and 75–85% aggregate/sand. Add more sand/aggregate or clay to achieve this, and to make a solid house that will last for hundreds of years.

Below: A clay-rich subsoil will roll into a sausage, and you should be able to wrap it around your finger without it breaking.

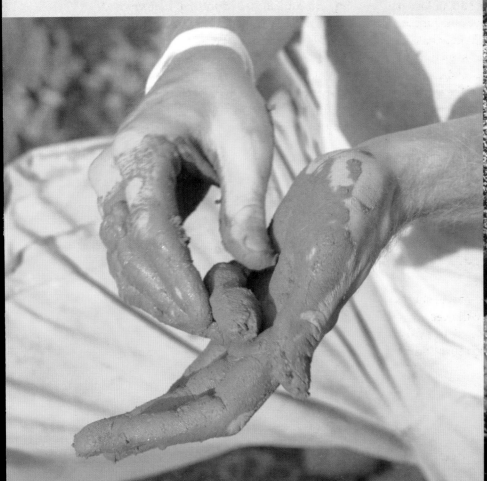

The ball test: Drop the ball from waist height onto the ground.

Too much sand and the ball will crumble: add more clay.

Too much clay and the ball will flatten into a pancake: add more sand.

Testing a dry-run cob mix

Once you have analysed your soil it is a good idea to make a few test batch mixes with the relevant adjustments made to your soil. There should be some variations within each test mix, i.e. different volumes of sands/aggregates added to reveal your ideal mix. It is very wise to find this out before you start building. To see how to make a cob mix please refer to Chapter 4: *How to make a cob mix*.

The brick test

Do not add straw to your mix for this test. This will test the amount of shrinkage to expect in your mix.

As described earlier, the more clay there is in the subsoil, the more it will shrink when it dries. A small amount of shrinkage is unavoidable, but excess amounts must be avoided. If your tests have revealed a very high clay content, you will want to add high amounts of sand/aggregate to balance this out. Similarly, if your soil is very high in sand, you may need some more clay to produce sufficient binding between the sand/aggregate particles.

1 Use a series of 250mm (10")-long wooden frames, at least 50mm (2") thick and 50mm (2") wide.

2 Fill with your damp test cob mixes, filling each box with a different mix ratio. Compact the mix well.

3 Allow each mix to dry thoroughly, then push all the material to one end of the box, and monitor your results: measure the total amount that the cob has shrunk away from the wood along the length. Multiply this amount by ten to find the percentage of shrinkage i.e. 0.30 x 10= 3% shrinkage. Anything more than a 4% shrinkage means that you need to add more sand to your mix.

If your bricks do not hold form at all and crumble apart, then you do not have enough clay in your soil.

The ball test (see above)

Using your dry-run cob mixes, pack a handful into a tight ball. The mix should be as dry as possible to avoid confusing results, as the water alone may hold the ball together. Drop the ball from waist height and analyse the form in which it lands:

- For an ideal mix, the ball will hold its form when it hits the ground.
- With a mix containing too much sand, the ball will crumble and fall apart on the ground. This means that it needs more clay.
- With a mix containing too much clay, the ball will flatten into a pancake. Add more sand.

If you are lucky enough to find a suitable clay subsoil in your own back garden or building plot, plan ahead with what you intend to do with the excavation hole. Think about how you can make it work positively for you and the landscape, instead of leaving behind an ugly and dangerous eyesore. Integrate this aspect of your build into the bigger picture of the design of your site: use the turfs for a grass roof on your building or on a shed, set aside the topsoil to be used for creating garden beds, and turn the hole into a pond which can provide beauty, and accommodate ducks, which can provide eggs for you to eat, fertiliser for the garden, pest and weed control, and so on.

“ Soil has a peculiar fascination, which impinges upon all of us at some time or other. Farmers or horticulturalists till it, engineers move it about in juggernaut-like machines, small boys (and girls!) dig in it, and mothers abhor it as being dirty . . . The soil has a vital and important role to play in the life of the world and mankind. ”

E. M. Bridges, *World Soils*

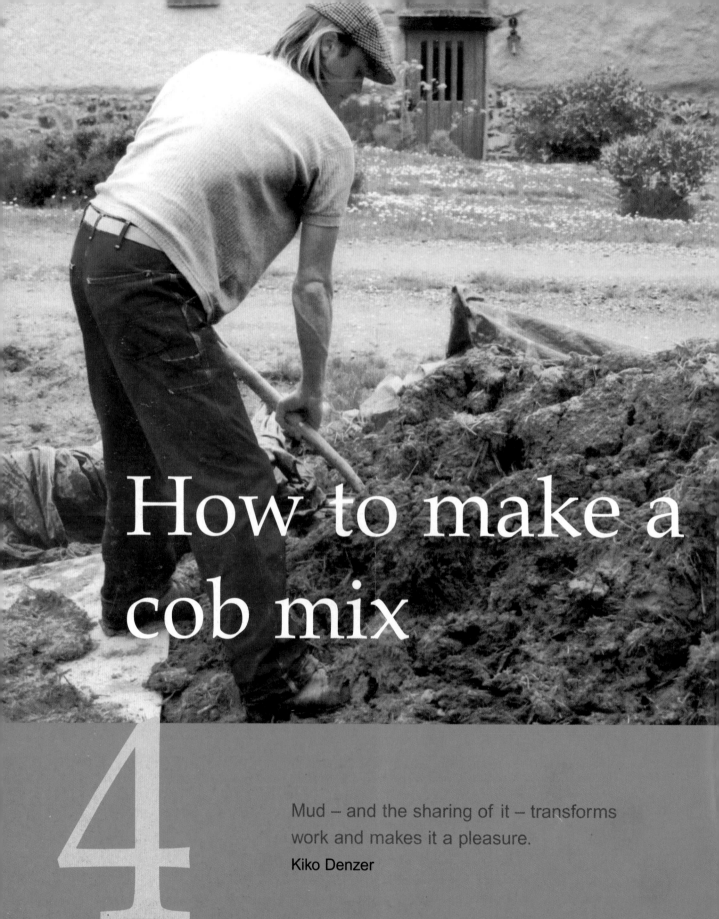

How to make a
cob mix

4

Mud – and the sharing of it – transforms
work and makes it a pleasure.
Kiko Denzer

Chapter 4

There are many different ways to mix up cob. From the very basic low-tech, low-impact, where the only piece of equipment you will need is your feet, hands and a tarpaulin (tarp), to the more high-tech, high-impact method of using a JCB or a tractor. After reading this chapter you should be in a position to choose which method is right for your job. Bear in mind that there is no single right or wrong way to do it; only the way that is right for your specific site, the tools you have available, your finances, your available manpower, and the size of your project. The following is a breakdown of all the tried and tested methods throughout history up to the present.

As well as the above, in this chapter we will talk about what tools and supplies you will need, give a breakdown of the basic ingredients to make a good cob mix: what their role is in the mix, what proportions to use them in, and the best places to source them. Don't be afraid to try out your own methods, and be open to experimentation.

Left: Pulling from a freshly made-up batch of cob. Middle: Mixing cob with horses. Right: Making cob with a JCB.

Clay subsoil and fresh long, strong straw

Aggregate / sand and water

Ingredients

To make a good cob mix you must have good ingredients. Your four basic ingredients are:

1 A suitable clay subsoil
   (discussed in Chapter 3)
2 An aggregate/sand
3 Fibre: fresh straw (not hay)
4 Water

Simple as that!

These materials can all be found in abundance throughout the UK, and you should try to source them as locally as possible to avoid unnecessary transportation,

and to create a sense of satisfaction that your building is being created from the elements of the land around you.

The functions of each material within the mix are as follows:

Clay subsoil

The function of the clay is as a binder to hold the aggregate particles together. Clay holds this aggregate mechanically, via suction. It must initially be made wet to coat the aggregates. Clay expands when wet and contracts when dry, and hence needs the addition of the fibres and aggregate to make it more stable, and to prevent serious cracking which could produce a weakness in the wall. This is why your proportions of aggregate to clay should be around 80% aggregate to 20% clay. Remember that you may find naturally occurring aggregate in your clay subsoil, and may therefore not need to add any to the mix. With this in mind it should help you to re-frame the notion of clay being the predominant ingredient of cob. It is not. The aggregate is your main ingredient, and the clay simply acts like a mortar between bricks holding the aggregate in one unified mass. In Cornwall, we have found that there is a layer of shaley stone called 'rab' just above the subsoil, which works as a perfect aggregate when mixed in with the clay. However, in most circumstances, it is going to be essential that you introduce an outside source of aggregate.

Aggregates
The purpose of the aggregate is to stabilise the clay, which as mentioned above is an inherently unstable material. The best cob is made by having a range of individual particle sizes, from fist-sized stone to gravel and sand to fine silts that are found naturally in the clay. Like a good mortar mix, the more sharp and angular the particles are, the better the results will be.

This ensures that the particles lock together and hence create a good bond, as opposed to rounded particles like beach sand, which will not. (To visualise this, think of how much more stable a stacked pile of bricks is with angular surfaces, compared with a pile of tennis balls.)

We get our aggregate (between 100mm/4" diameter and dust) from a local quarry, delivered to where we are mixing the cob.

This produces a very good mix. Never use beach sand, as the particle sizes are rounded and would be paramount to trying to stack the pile of tennis balls mentioned above.

Fibres
Traditionally, people have used all sorts of different fibres that they have had to hand, such as heather, rushes and flax. The best fibre is fresh, strong, long straw. Oat, wheat and rye are the best options. It must be either very fresh or have been well stored, away from moisture. Do not use straw that has got wet and begun to rot; its function within the wall will be futile. You need strong straw that is not brittle and does not break easily.

The purpose of the fibre is many-fold. Firstly, it gives tensile and sheer strength to the wall, acting as a natural re-bar. This allows the cob wall to move slightly and therefore withstand ground settlement and natural movement within the building. Secondly, it prevents major cracking, by spreading out the cracks as hairline fractures, thus preventing the development of larger, structural cracks within the wall. Thirdly, it can soak up excess water during the mixing process. Finally, it adds insulation to a cob wall by trapping air in its hollow stems.

A question that is often asked is whether the straw rots out inside the wall. The answer is that it rots out very little, because there is little oxygen or moisture available once the wall has dried out. We have pulled apart cob walls that are a couple of hundred years old and found much of the fibre to be intact.

The best way to get straw is to source it direct from a local farmer or from a farm supply store. Straw will always be cheapest when it is in abundant supply, and therefore the best time to purchase it is immediately after it has been harvested – in mid-summer and early autumn. This way you can also ensure that it is going to be as fresh as possible. If you do have to purchase it at another time of year, make sure that what you are buying has been stored well and shows no sign of being damp and rotten.

Water

Nothing on earth is possible without water. The addition of water to the above ingredients will transform the clay, straw and aggregate into the thick, sticky, homogeneous material that is going to make up your walls. As mentioned previously, the binder, i.e. the clay, must initially be made wet to coat the aggregate particles to create the suction that sticks it all together.

There is not much to say about water apart from how much or how little you want to add to your ingredients. Too much, and the clay element becomes unmanageable; too little, and the adequate bonding will not take

Tools you will need for mixing using the tarp method

1 Tarpaulins (plastic woven sheeting purchased from a builders'-merchant) – anywhere from 2x2.5 metres (6'6"x8'3") to 2.5x3.5 metres (8'3"x11'6")
2 Shovels and forks to transfer material from pile to bucket
3 Buckets to measure out proportions of material
4 Water source with hose if necessary
5 Sturdy feet (with or without boots) for stomping
6 Good friends to stomp, mix and have fun with

place and a homogeneous mixing of all the ingredients will not be achieved. We will clarify how much water, and how to know when you have added enough, when we talk about mixing later in the chapter. It goes without saying that the water you use should be as clean as possible because you will be touching every bit of cob that you mix up. Get some catchment water tubs next to where you will be building, to utilise the rainfall.

Just a note to remember: the above information on the separate ingredients will help you to make the best cob you can. There have been times when due to circumstances we have made cob using poor ingredients – any old sand we had to hand, and even wet straw, and we were still able to make a totally adequate cob mix. We want to reduce any anxiety you may have about having to have everything absolutely perfect; cob is a very flexible and forgiving material. It is like making a loaf of bread: you can use poorer quality ingredients, and still come up with something that looks and tastes like bread. It just may not be the best bread you have ever tasted!

The two main methods that we use are the mixing by foot tarp method and mixing with a JCB.

Mixing cob using the tarp method

1. Add proportions of dry clay and aggregate/sand onto tarp.

2. With partner, fold tarp from side to side to mix dry ingredients.

3. Add water to dry ingredients to moisten the mix.

4. Stomp till ingredients are thoroughly mixed, fold tarp and stomp again.

5. Add the straw and stomp until each strand is coated in clay.

6. When thoroughly mixed, fold tarp to form a large 'burrito' of mixed cob ready to go.

Proportions of clay subsoil and sand/aggregate should be ascertained through soil tests explained in Chapter 3, and should roughly follow the 15-25% clay subsoil to 75-85% sand/aggregate rule.

You can create a crater in the middle of the dry ingredients into which the water can be added, and then fold the dry ingredients in on top. Don't add too much water too quickly. If you end up with a really sloppy, squelchy mix you may need to add more dry ingredients to firm it up a bit.

When folding the tarp always bend your knees to go easy on your back. Keep loads to within a weight that you and your co-workers can manage comfortably.

Have fun whilst stomping your cob. Use a variety of moves: 'the twist', to really smear the clay particles into the sand/aggregate; 'the high knee': to get a good downward thrust into the mud. Enjoy yourself!

The tarp method

The tarp method was developed in 1994 by a cobber in North America called Becky Bee. This method is best carried out with a partner, but can also be done solo. If done properly, we have found that it is much kinder on the body than the traditional method of turning the mix over using a fork and spade (described later in this chapter).

There will be lots of foot stomping involved in this method, so the best footwear to use is wellingtons or other boots you don't mind getting very muddy. Some people prefer to wear something with a less deep tread because you can end up getting literally stuck in the mud! Where we learned in North America with the Cob Cottage Company, we wore the best footwear available – nothing – and developed hard, leathery soles to protect our feet. However, the clay that we were using had no sharp stones in comparison to the shaley cob that we tread in the UK. This method did not translate well. If your cob is not stony, try out mixing with bare feet – it feels wonderful! This tarp method can be a very joyful part of the process of building your own cob creation. It is the best place to bring in the whole family – kids, grannies, stressed out executives, etc – to have fun and get some good fresh air and exercise, all the while making the material you will be building your house with. You will be amazed at how much fun hard work can be.

Refer to the six steps of making a cob mix on page 56, and use the following explanation to bring more clarity to the pictures. For the tarp method roughly 4–5 five-gallon buckets of material is a good amount for two people to handle.

1 Get all materials ready, close to your mixing space i.e. your clay, sand/ aggregate, straw and water, including five-gallon buckets.

2 Take your tarp and lay it out on a flat space. You can have as many tarps going as you have space and people, to get higher volumes mixed up quickly.

3 Place dry ingredients – clay subsoil, sand/aggregate (if your clay requires you to add extra) – in a pile in the middle of your tarp. Refer to the ratios that you analysed through the tests described earlier in Chapter 3. There are no standard ratios that we can give you here, because every clay subsoil differs from place to place. As a guideline, refer back to the 80%/20% aggregate to clay rule which seems to be the standard ratio for most of the old cob houses built in the south-west of England.

4 If you have had to add extra aggregate to your clay subsoil, now is the time to mix these two dry ingredients together before adding water. Mix by having two people each grabbing one corner of the same side of the tarp. Walk forward so that the tarp is folded in half and the dry material is

deposited at one end. Return the tarp back to its starting position and lay it flat on the ground. Now walk to the other end and grab those corners and repeat the process, flipping the dry material to the other end of the tarp. Do this roughly four times or until your dry ingredients are thoroughly mixed, at which point you should fold the tarp and then lay it flat so that your dry ingredients end up back in the centre.

5 Make a crater in the middle of your dry ingredients, and add a small amount of water into the crater. Always err on the side of caution when you add your water. It is much easier to add more than it is to take it away. A standard ratio of five parts dry ingredient to one part water is usually good, but your mix may need a bit more or a bit less. You will develop this judgement the more you get used to mixing cob. Keep one thing in mind about adding too much water: even though it may feel easier to mix, a mix that is too wet will not hold its form well once you start to build, and the wetter your cob mix, the more liable it is to crack when it starts to dry out. If you do end up adding too much water, either leave the batch to dry in the sun, or you can add a bit more of the dry ingredients. The straw will soak up some of the excess water as well.

6 Push the dry ingredients with your feet into the middle of the crater of water, and begin to stomp. Hopefully this is what you have been waiting to do ever since the idea of building with cob was sown in your mind. Grab as many humans as you can fit onto the tarp without bumping into one another, and twist and shout and shake it all about. Jogging on the spot, twisting your heels, dancing, whatever feels good just so long as the end result is that all water and dry material is mixed together, and the clay particles and aggregate smear together.

7 When this has been achieved, grab your tarp corners again and fold the tarp back on itself a few times then stomp again. If you are having difficulty mixing the dry ingredients into a homogeneous mix, add some more water until this is achieved. By now your cob mix should be forming itself into the shape of a huge burrito every time you roll the tarp back and forth. This is a good indicator to let you know that the cob is well on its way to getting thoroughly mixed. Another test you can do is to roll the cob into a ball in your hand: it should hold form like plasticine. You are now ready to add the straw.

8 Stomp the mix out of its burrito form so that it spreads across the tarp. Grab a few handfuls of straw, and sprinkle evenly across the surface of the cob. Stomp vigorously until the all the straw has been smeared by cob. Turn the mix over again and stomp the mix flat. Add more straw and do the same as above. A good ratio of straw to clay and aggregate is one five-gallon bucket of compressed straw to five buckets of dry ingredients. Keep repeating this process until the straw is used up, constantly turning and stomping, until all the straw is thoroughly mixed in and is the same colour as the cob. Add straw until the mix can take no more.

Top left: Placing the dry ingredients on the tarp.

Top right: The clay subsoil and sand mix ready to turn.

Far left: Adding water into the crater.

Left: Folding tarp over as a one-man band.

Below: Adding straw while stomping.

Bottom left: The merry dance.

Bottom right: The fruits of your labour – a pile of cob ready for building.

By now you should have a large burrito of very straw-rich cob sitting in the middle of your tarp. Grasp a breath of air and get ready to build!

Mixing by foot is an excellent method as long as you have time and people at hand. This method is very complementary to the ecological philosophy behind green building practices, and will satisfy you if you are adamant about not having any heavy machinery involved, and not wasting precious fossil fuels. It is also excellent if logistically you cannot get heavy machinery on to your site. It will create little or no damage to the surrounding ecosystem where you are building.

Mixing by foot without tarps

You will need:
1 Shovels and forks for turning cob over
2 A water source with hose

1 You will also need a plywood or concrete base that is relatively flat. This must be sprayed down with water before you begin.

2 Place your clay subsoil and aggregate (if using), in a pile and wet down.

3 Stomp the material and then turn with a shovel until mixed thoroughly.

4 Add straw as described on page 58. Turn and tread again until ready.

This method is very effective, but we have found it can be really hard on your back and quite a grind to make large mixes.

The pit method

You will need:
1 Shovels and buckets
2 Large tarps
3 Straw bales
4 A water source

This is a good method if you find yourself without any help, because you can achieve large mixes by yourself. It also provides an opportunity to soak the clay before mixing, which will make it much easier to work with – ideal if you are re-using dry cob from old walls. This is also a good method for young children who don't have the strength to roll a heavy tarp. It is like a playpen for kids to slosh around and mix in, and you will probably find most of the adults in there too.

1 A very simple and effective way of creating a pit to mix in is by placing four or more straw bales in a square with their corners touching.

2 Drape a large tarp over the hole in the middle, overlapping the bales. Do not attach the tarp to the bales because you will want to be able to manoeuvre it later. Instead, rest some heavy stones on top of the tarps to keep them in place.

3 Put your dry ingredients in the middle of the pit and enough water to get all your

dry ingredients wet and mixed well. You can pull the corners of the tarp from one side to the other rather like in the tarp mixing described on page 57, in order to bring the dry material from the edges into the middle.

4 Start stomping and have a good time playing in the mud.

5 Once all your ingredients are mixed evenly and thoroughly, start to add your straw as described in the tarp method.

6 Once the mix is made, roll your cob into balls by hand to transport to the wall, because using a pitchfork will inevitably lead to punctures in the tarp.

Using a cement mixer

This is our least favourite of all the mixing methods and seems like a lot of hard work to produce unsatisfactory results. However, we still want to share it with you so that you can make up your own mind. The main disadvantage is that to enable the ingredients to be mixed thoroughly it needs to be mixed very wet, almost slurry-like, which has negative implications for when you come to build. The single advantage for using a cement mixer for mixing up cob is that it allows for mixing to take place inside and in a smaller space. You need a revolving-drum cement mixer and it is necessary to reset the drum at a horizontal position and even

tipped forward slightly. When we have experimented with this method, we followed the lead created by English Heritage, who restored a large cob property at Bowhill near Exeter from the 1970s through to the 1990s. Like them, we found that the inherent problem with using a cement mixer is that the cob forms into balls which are dry on the inside and generally inadequately mixed. One way to overcome this is by making sure the mix is very wet. Tip the balls into a box placed beneath the mixer, then tread each individual ball to mix the wet outside with the dry inside. Depending on the consistency of the mix, you may be able to build with it straight away. But if they are too wet, as we found most of the time, it is best to leave them for a day or more to firm up. With so many other satisfactory options, unless you are really logistically challenged, this method is a whole lot of work for sub-standard results.

The JCB method

There are always compromises to be made in this world. We have found that on large-scale projects, using a JCB to mix our cob has been one such compromise. One of the pretexts of natural building is that it must be sustainable. Within this pretext we must consider how sustainable it is for our bodies and energy output. To ensure that we don't wear ourselves out physically, we have welcomed the use of a machine that can mix the same amount in five hours as it would take two people three days and a lot of stomping. If you decide to take this route, be mindful of the fossil fuels that you are burning, and also the destruction a JCB can cause to a site. There are two routes you can take here. You can either hire out a digger with its driver, and instruct him/her to make the mix while you add the straw and water. Or, you can hire out a digger and drive it yourself and get someone else to add the vital ingredients.

It is possible to use a large JCB digger, a smaller mini-digger, or a tractor. For the former, it is easiest to mix the cob *in situ* in the pit from where you are extracting the clay, as it has a long enough arm reach to access the far reaches of the pit. If using one of the latter two, a hard flat mixing area is needed – a cement pad is best – to enable you to use the treads of the machine to do the majority of the mixing.

1 Scrape away all topsoil and organic matter with the bucket creating an area roughly 3 metres (10') wide to 4.5–6 metres (15'-20') long.

2 Cut into the clay with the arm and claw of the machine and break it up until you have created a pit of broken-up clay at least 2.5–3m (8'-10') deep. Continue turning the clay over.

3 Turn on the water source. It is best to have a long hose, and to dangle it into the side of the pit so that it does not get in the way. This water can be left in one place for about an hour. You can also fill the machine claw and distribute it where needed. You may want to change the hose to the other side of the pit halfway through.

4 Add aggregate if necessary, and mix with the clay and water. Continue turning the mix over with the claw for roughly an hour, or until the mix starts to form a sticky cohesive mass.

5 You are now ready to add straw. Stand on the side of the pit and chuck in armfuls of straw. You will need in total roughly one bale per tonne. Continue to turn over with the claw until all the straw is thoroughly coated in clay. Take care to monitor your water content and make sure it doesn't get too sloppy. After roughly five hours of mixing, the JCB should have mixed up about 12 tonnes of cob.

6 Prepare a hard plywood base or other flat, smooth, clean surface next to the spot where you are about to build, and transport the freshly mixed cob to this area. This fresh cob must be covered with tarps, to prevent it from drying out in the sun or from getting too wet in the rain.

The mixing practice when using a mini-digger or tractor on a splat pad is similar to the above method.

The main difference is that you can use the tyre treads to drive over the cob instead of turning it over so much. Every fourth time you roll over the mix you should use your JCB bucket to turn it over.

Right: Turning the mix with a mini-digger.

Below: Excavating subsoil with a JCB.

Above: Delivering subsoil to the mixing area.

Above: Adding straw during the mixing stage.

Left: Delivering a trailer-load of ready-mixed cob.

Some cob builders such as Kevin McCabe in Devon take machine cob one step further, and actually feed the cob onto the wall with the machine bucket and then tread it into place. This is the ultimate in time-saving and efficiency when building long stretches of wall.

to a time when people were in the habit of finding ways to minimise their already stretched energy output. This is the traditional version of using a digger or tractor to make your mix, minus the fossil fuels.

You need a contained pen or yard into

> Traditionally, mixing was a process to be learned, with judgement developing out of working directly with the wet material, since the nature and requirements of the particular soil being prepared become apparent as mixing proceeds.

Ray Harrison: *Bowhill, English Heritage Research Transactions*

One disadvantage of mixing up such a large amount of cob in one go (and it is good to mix up a large batch to maximise the use of the machine) is that you may not be able to build quickly enough before the cob dries out, especially in mid-summer. If your cob does become hard and unworkable, you may need to add more water and rework the cob with your feet. It is ideal to use as fresh a cob mix as possible, to avoid having to mix it twice, as well as keeping the straw fresh. Good practice is not to have your cob sitting around for more than three weeks before being used.

which your clay, straw and water are scattered. Add your animals, and leave them to trample the raw ingredients over night. Your mix should be ready by the morning with the addition of dung for extra fibre and stickiness!

How to know when your mix is ready

This is something that you will develop by trial and error. In the meantime, there are some simple tests that you can do to find out whether your mix is ready to go and good to build with. On the whole, it should be thoroughly mixed, and you should not see any dry bits or be able to distinguish any aggregate/sand within the mix. Similarly, the

Mixing with animals

This method of using animals – cows, horses or bullocks were most often used – dates back straw should be thoroughly incorporated and well dirtied by the clay. It should not be sloppy, nor too stiff. It should hold its form well. Grab a large handful, and with a partner try to rip the cob apart. If it is hard to do this, you know that you have enough straw in it, and that the straw will be able to serve the purposes for which it is intended. Refer back to Chapter 3 for more tests, which will help you to know if your mix is ready, such as the ball test (see page 48).

"Traditionally, mixing was a process to be learned, with judgement developing out of working directly with the wet material, since the nature and requirements of the particular soil being prepared become apparent as mixing proceeds. As no two soils are identical in their constituents, so each mixing will vary slightly from the next. Variation will be in the amount of water and fibre needed and possibly in the need to add extra aggregate. In these circumstances taking things gradually by adding water and fibre by degrees makes common sense and is the recommended practice for those starting to learn the techniques. Once the basic management of the material is understood through experience, then shortcuts can be considered." (Harrison, Ray: *Bowhill, English Heritage Research Transactions*)

Clare Marsh putting her horses to work mixing cob. Higher Boden, Manaccan, UK.

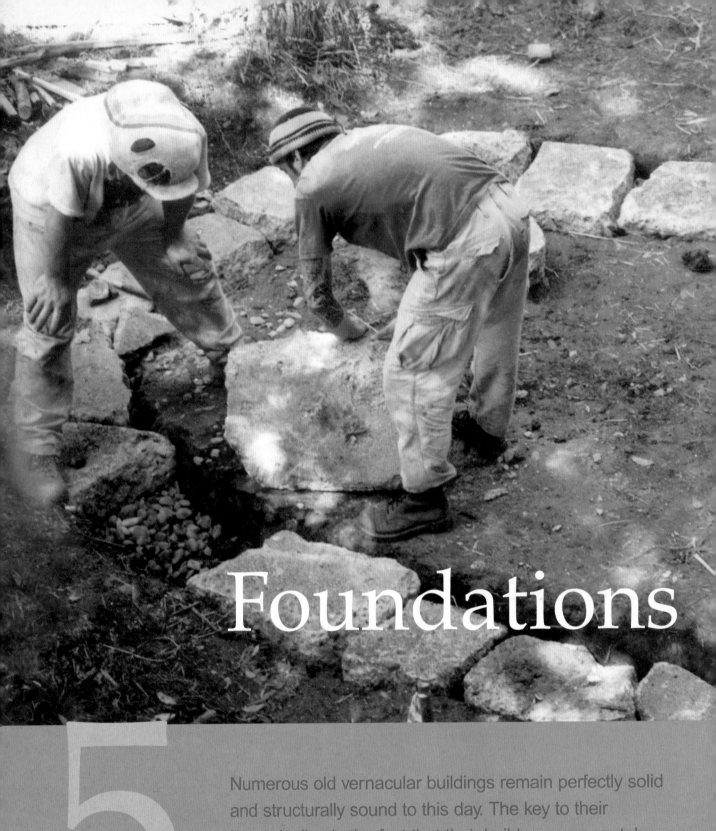

Foundations

5

Numerous old vernacular buildings remain perfectly solid
and structurally sound to this day. The key to their
longevity lies in the fact that their builders were acutely
aware of the lie of the land, the local climate and the
components of the ground on which they were building.

Chapter 5

The foundations of a building serve the purpose of carrying the weight of the load of the building: the roof, walls, furniture, floors, and the people inside. The foundation distributes this weight evenly over a wide area of solid ground so that the building is prevented from sinking into the ground, potentially causing cracking in the walls, and even possible collapse.

The foundations for a cob building include the below-ground works and the above-ground stem wall or foundation plinth on which the cob sits.

The below-ground works, or footings, support the load below ground level, evenly distributing its weight to the ground beneath it. The stem wall, or foundation plinth, is built above ground at least 450mm (1 1/2 feet) and serves to protect the walls and floors from water damage – both the splashback from rain hitting the ground, and the damage that would be incurred if the walls were to come into direct contact with water on the ground. This is especially important for cob walls, because otherwise they would literally turn to mush and be washed away after the first season of rainfall.

Opposite: Laying a foundation out of 'urbanite' (recycled pieces of concrete). Below left: Digging the drainage trench for the foundation. Below right: Cob walls must rest on a stone plinth at least 450mm (18") off the ground to protect from moisture.

Traditionally, houses were built directly onto the solid earth beneath them. The topsoil was scraped off and the ground dug until a stable base was reached. The foundation stones would have been laid onto this solid base below ground level, built wider at the base to disperse the load of the building above. As ground level was reached, the stones were gradually narrowed and built at a width to accommodate the width of the walls built onto them.

All over the UK and Ireland, hundreds of thousands of ancient, vernacular buildings built on these sorts of foundations remain, and are still perfectly solid and structurally sound. The key to their success is that they were sited correctly. That is, the builders were acutely aware of the lie of the land, the local climate, the components of the ground on which they were building, and how this related to the kind of building that they were creating, e.g. heavy cob walls, or a light timber frame. For example, solid bedrock will carry more weight than soft clay. These remaining buildings were obviously built where the ground was well drained, and provided a stable, uniform sub-base. The use of simple land drains around the perimeter of the building would have been common practice to move water quickly away from the base of the walls. These practices can be used successfully today to provide simple solutions for supporting your building.

Modern construction methods generally rely on using vast amounts of cement to create artificial strip and slab foundations. This has become the accepted norm due to two main reasons: houses are built more often in areas where the subsoil and drainage are not suitable for construction, and the use of cement allows us to modify the land to achieve acceptable conditions. Also, people seem to have lost confidence in the ability of natural, simple methods to provide effective and more than adequate solutions. Most modern buildings are built to standards that cater for the worst possible case scenarios instead of being site-specific.

Site-specific drainage and foundation requirements

Below are some questions to ask yourself before you decide what foundations you will need.

1 What sort of roof/wall loads do the ground and foundations need to accommodate? A thatch roof is heavier than a slate or shingle roof, and cob walls are heavier than a light timber-framed building.

2 What is the capacity of the natural ground to take the weight of the building above it? Compacted bedrock will take more weight than soft clays.

3 Which system will be easiest to build and will come within your available financial means?

4 Are the majority of the materials able to be sourced locally (i.e. stone, sands, lime putty)? Consider that many hydraulic limes are imported from France and the Continent.

5 How deep does the ground freeze in the winter, if at all?

6 What are the local drainage conditions on the site, and what is the lie of the land?

For a modern, newly-built cob building we recommend two different types of foundations that will eradicate the need for cement, will be simple to construct, and will be suitable for the range of climates across the UK and Ireland: a rubble trench foundation with a local stone and lime stem wall, and a lime-crete strip foundation with stone plinth and surrounding curtain or land drain.

Rubble trench foundation

A rubble trench foundation is a very simple and effective method to establish a lasting foundation, which incorporates drainage and load-bearing capacities all in one elegant system.

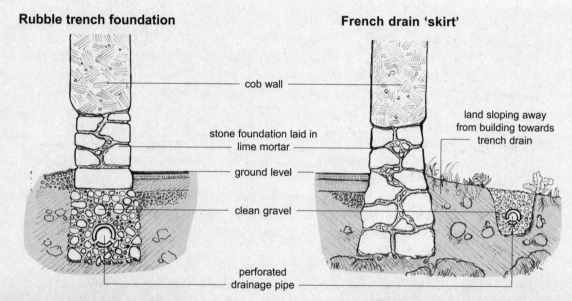

Rubble trench foundation

French drain 'skirt'

cob wall

stone foundation laid in lime mortar

ground level

clean gravel

perforated drainage pipe

land sloping away from building towards trench drain

Bottom left and right: The stone stem wall with the first layer of fresh cob on top, showing 'drip lip' (see page 87).

Above left and right: Two different options for foundation systems incorporating drainage, suitable for cob buildings.

Left: Clearing the site and marking out the building's outline. Right: Digging a trench for drainage.

How to lay a rubble trench foundation system

1 Clear the site by removing all vegetation, topsoil and organic matter.

2 Mark out the perimeter of your building using pegs and string or paint.

3 Dig a trench slightly wider than the width of the stem wall down to solid, stable ground. This involves removing all the topsoil, all organic matter and plant roots. For a 2' (600mm)-thick cob wall, make a 2' 6" (750mm)-wide trench, with 3" (75mm) extra on both the interior and exterior of the trench.

The bottom of the trench should be sloping down at all points towards one exit point away from the building – roughly 25mm for every 1.5m (1" for every 5') as a guide. To test this, dump a few buckets of water into the trench and make sure that the water runs instantly down towards the exit point and away from the building. Ensure that the

walls of the trench are fairly vertical and that the floor is clean and well compacted. In very cold climates where the ground freezes for part of the year, it is necessary to consider where the frost line is, as this will dictate the depth of the foundation trench. The frost line demarcates the lowest point at which the ground freezes, and any trench will need to be dug below this level. This should not be an issue in most areas of the UK, but on high land and in some parts of Scotland you may need to pay heed to this.

If you start digging up large tree roots from a tree nearby when you are digging your foundation trench, you may need to think about whether it is a suitable site for building. Large tree roots can have the strength to upheave a building. Utility ducts can be laid in the bottom of the rubble trench before backfilling, so that electrical cables, water pipes, and phone lines can be inserted at a later date.

4 Half backfill the trench. Start backfilling the base of the trench with clean drainage

70

rock or gravel, 25–100mm (1"–4") in diameter, roughly 150mm (6") deep. Thoroughly compact this rock using a tamper – mechanical or home-made (see Chapter 10, page 166).

5 Into the compacted rock trench lay a 100mm (4") diameter perforated drainage pipe that circles the whole perimeter of the building and extends at least 3.5 metres (11' 6") beyond the exit point away from the building, so that the draining water will come to the surface at a point away from the building.

If the land does not naturally slope away from the building, it is good practice to create a well (about 1.2 metres/4' deep) filled with rubble, or stones (if purchased, these will be very expensive, so see what you can find around you) that the pipe can drain into. Ensure that the pipe follows the contour of the slope that has been created in the trench.

6 Backfill the rest of the trench with your drainage filler in roughly 150mm

(6") increments, tamping successively each layer until you reach 150mm (6") below ground level. Set down a layer of hessian sacking or a similar open-weave fabric on top of the drainage filler, to prevent silt and soil from clogging the rubble and hence impeding its drainage function.

7 Begin laying your first stones for the stem wall. It is good practice to use large stones for this, and they should be as flat as possible for extra stability, as these provide the base on which the rest of the plinth, the cob walls and the roof will sit.

These stones should span the full width of the trench and should be bedded into the gravel very thoroughly. If there are any wobbles, these should be stabilised by chipping away protruding pieces of stone or by using small pieces of stone (chinkers) to wedge between the loose areas. The stones should feel absolutely solid beneath your feet as you walk on them. These base stones should come up roughly to ground level.

Above: Use a 'dyke-hammer' to shape the stone for a close fit.

Below: The stone plinth should be constructed to contain a cavity for an insulative filler. There should be 'tie' stones at intervals linking the two walls together.

8 Now build up your stem wall. The next stones laid should narrow to the actual width of the cob wall. From this point you should start bedding the stones into a lime/sand mortar (see recipe below opposite).

Basic stone-laying principles

1 Pick out as many stones with good flat faces as possible. A good stone will have at least one good flat face with a ninety degree angle; however, most stones can be worked with (it just takes a bit more fiddling around!)

2 Lay your stone selection out on the ground next to the wall you are building, and take some time to familiarise yourself with the stones, so that over time you can quickly select stones for a specific fit without having to search too hard.

3 Always use your largest stones on the bottom and get progressively smaller the higher you build up. Not only does this minimise lifting the heavy stones, but also means that there is a larger surface area on the bottom layers to evenly distribute the load above.

4 The lime mortar should be wet enough to allow good suction between stones, but not so wet that it slops everywhere, creating messy stones and inadequate support between the stones. The real purpose of the mortar is as a bed to cushion in between each stone joint.

5 Stones should be fitted together solidly so that if there were no mortar they would still be stable. The use of mortar will not replace good stonework.

6 Always stagger the stone joints and never create vertical joints, which would be liable to vertical separation. Stones should be laid so that the joint between every neighbouring stone is bridged by the stone above it.

7 Create two separate walls with a cavity in between, and use large stones at intervals to tie the two walls together. This cavity can be filled with an insulative material such as expanded clay balls (e.g. 'Opti-roc'), perlite, vermiculite, or clean rubble. This is an essential area to insulate in a cob house, in order to help to increase the overall thermal efficiency of the walls.

Left: Lime mortar fill in the stone plinth ready for the first lift of cob.

Right: Bricks or blocks can be used to build the plinth instead of stone.

Lime/sand mortar mix:
1 part lime putty : 4 parts coarse sand : 1/4 part pozzolanic additive

Pozzolan facilitates the setting of lime putty deep inside a wall, as putty needs contact with CO_2 in the air to set. Pozzolans can be purchased from lime suppliers or you can use purchased or home-made brick dust! You can also substitute hydraulic lime for the lime putty. Use NHL 5, which is good for damp conditions and will not need pozzolan.

Lime/sand pointing mix
1 part lime putty : 3 parts coarse sand

Lime-crete mix
1 part NHL 5 to 3 parts aggregate

recipes

8 Each stone should sit solidly into the bed of mortar – if it wobbles, use a masonry chisel and a lump hammer or a dyke hammer to remove any extruding bumps. You can also use small shim stones (it is good to have a large pile of these at hand) to wedge under a stone to help stabilise it. If it still does not sit well, select another stone. There is an old stonemason's saying that states: if a stone doesn't fit after trying it in three directions, find another stone!

9 As you build up the stone plinth, it is essential to create plumb, vertical faces on both the inside and outside faces of the wall. Use a level to make sure this is achieved.

10 The final layer of stone should be level width-ways, and not sloping toward the outer and inner faces. The latter may cause the cob to slip off the stem wall. The stones do not need to be completely level horizontally however, as some undulations will provide a good key for the first lift of cob to anchor into.

11 The mortar in the stone stem wall should have completely gone off before the first lift of cob is laid; although setting conditions will vary, it is good to wait at least ten days before building.

Why use lime mortars with natural stone?

In medieval times ". . . The preparation of mortar was one of the trades valued equally alongside stonemasonry and carpentry. The master mortar-men would assess the locally available materials and blend and work them to suit the requirements of the particular building, judging the nature of the stones to be bonded, the climate likely to influence the building and the particular part of the building.

Stone joints should be pointed with a soft lime mortar to protect the stones from erosion and for aesthetic sensibilities.

The pointing mortar should be brushed stiffly with a churn brush before it has completely set.

They arranged for their men to beat and chop the mortar to get the very best workability and economy of materials. All this is in very sharp contrast to the organisation of a modern building site, where very little care and attention is taken in the preparation of mortar. We have come to expect the strength of cement binders to compensate for the use of poor aggregates

- Soft lime mortars will accommodate the inevitable moving and shifting of a building, whereas brittle cement is more likely to crack with these subtle movements.
- Traditional building methods such as cob walling do not require strong mortars in their foundations because their shape, density and thickness provide com-

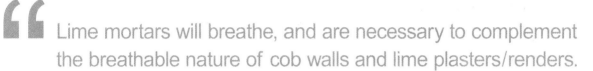

"Lime mortars will breathe, and are necessary to complement the breathable nature of cob walls and lime plasters/renders."

and for poor workmanship, but to get the best out of lime, and the best out of masonry, this attitude is not acceptable." (Stafford Holmes and Michael Wingate, *Building with lime: a practical introduction*).

There are many reasons why lime mortars should be favoured over cement as a bedding and jointing material for use with natural stone or brick:

- Lime mortars (especially those made with putty) remain 'soft' (yet thoroughly durable) even when fully set. This means that stones and bricks can be used again and again over time, as the lime will allow the masonry units to be easily disassembled without causing them damage.

pressional stresses on the loads below, which means that sheer gravity weighs everything down and ties everything together. Consider that many surviving ancient stone and cob buildings relied on dry-stack stone foundations or earth as a mortar – this is amazing, even when taking into account that there were many excellent stonemasons.
- Lime mortars will breathe, and are necessary to complement the breathable nature of cob walls and lime plasters/renders.
- Lime mortars (lime putty) can be re-used.
- Cement used in between natural stone can produce salt crystals, which can damage stone or brick.
- The pointing mortar used in between the

joints of a stone or brick should always be softer than the masonry. Its purpose is as a sacrificial element, so that under natural stresses such as rain and wind, the jointing material takes the brunt of the damage, eroding before the masonry. The idea is that the masonry outlives the exposed jointing mortar. The general rule is that the mortar should always be softer than the masonry, the latter being the more valuable.

When to use hydraulic lime and when to use lime putty

It is really only necessary to use hydraulic limes when the area you are constructing is going to be subject to continuous wetting or dampness. This is why it has traditionally been used in the construction of harbour walls, drainage, water storage tanks and sub-structures in damp ground.

We always recommend using NHL 5 (the strongest lime available) for strip foundations. For all other purposes lime putty is preferred, although the use of pozzolans (see lime mortar recipes section on page 73) may be needed to achieve a stronger set when the putty will not receive adequate contact with the air to achieve its full set. For more information on lime, see Chapter 9 on lime and other natural finishes. Stone needs to breathe to remain healthy, just like cob, although to a lesser degree. Lime should be used within an overall breathing system from the foundations up. As the stem wall is acting as a capillary break, it is important for any rising damp to be able to escape easily through the joints in

the masonry. For this reason neither type of foundation involves the use of a plastic damp-proof course, as is the standard for most modern constructions. Non-breathable, waterproof membranes are the antithesis to natural, breathable, earth wall systems.

A cob building that is sited correctly and has a self-draining foundation system (e.g. rubble trench) and the benefits of breathable finishes such as lime and earth, will effectively self-regulate any moisture that enters the fabric of the building. If a plastic membrane is used, the natural movement of capillary moisture will be restricted, and may cause damp problems in the future.

Pointing the stem wall (see page 74)

The exposed joints in the stone wall should be pointed with a lime mortar to protect the stones from erosion (it provides a soft, sacrificial element so that the mortar is eroded before the stone) and to make the stonework look neat and tidy.

The pointing should be carried out with a lime/sand mix (see the recipe box on page 73).
- A special pointing trowel should be used to tuck the mortar tightly into the joints.
- Apply the lime mix as dry as possible.
- Wet the joints prior to application.
- Allow the mortar to go off 'green hard' before using a stiff churn brush to smooth out any trowel marks and to compact the lime into the joint.

Stone foundations – the variations are endless!

- Clean any lime residue off the stones immediately with water and a stiff brush. If the residue dries, a wire brush should do the trick.

This foundation system is good when there is a supply of local stone which is not too expensive. Recycled stone can also be used. Stone is the perfect complement to a cob wall, and will look very beautiful.

It is, however, a very labour-intensive method, and is slower than using standard building blocks such as brick or concrete blocks. It can be expensive using a professional stonemason. If using bricks or concrete blocks, the same steps will also apply.

As an alternative to the above (see page 68), broken up bits of concrete ('urbanite') make excellent stacking units for a solid foundation. This makes use of a recycled material, is cheap, and can actually look quite attractive if done with care. It can also be rendered over. Check with your local county council to find out whether they are breaking up driveways or pavements near you, or where construction workers may be demolishing areas of concrete to make way for new developments. As seen on page 69, an alternative but similar foundation system involves placing the drainage element around the building, so that it falls under the eaves of the roof. This is known as a trench drain or land drain, and is suitable for use with a thatched building.

Lime-crete strip foundation

As an alternative to the conventional concrete strip foundation, the same technique can be applied, but replacing the cement with a strong hydraulic lime (NHL 5). This mixture of lime and aggregate is called 'lime-crete'. Although more expensive than cement, the use of lime will be kinder on the environment and will be in sympathy with the rest of the breathable materials making up your building. Lime-crete has been used for foundations for the last 2,000 years. You may be surprised to learn that some of our most well-known and lavish structures in the UK sit on 'lime-crete' foundations, such as the Houses of Parliament, the British Museum and Westminster Bridge. These should be proof enough that lime has adequate compressional strength to take a humble cob structure.

The strongest hydraulic lime (NHL 5) is used because of its direct contact with the ground, and hence its potential need to withstand prolonged dampness (although if properly drained, this should never be a problem). A land drain or curtain drain is used around the entire perimeter of the building, situated so that it can capture the water running off the roof eaves (if the building is going to be thatched). The trench should be roughly 450mm ($1^1/2$ feet) wide at the top, and can be narrowed towards the bottom so that it creates a v-shape. The land drain is constructed in exactly the same way as the rubble-trench drain described on page 69. The only difference is that you are building the drain on the outskirts of the building and not directly underneath it. This will serve the same purpose – to move water quickly away from the building (see diagram on page 69).

with clean gravel, tamped very thoroughly, and the lime poured on top of this. Ensure that the foundations are thoroughly set before building your foundation plinth, and bear in mind that lime-crete will take longer than concrete to set fully.

Left: Recycled broken up pieces of concrete ('urbanite') can make for attractive, solid and easy to build foundation stem walls, as the foundation for the cob bench shown left demonstrates. *Cob Cottage Company.*

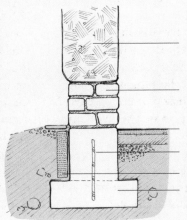

Poured lime-crete footing

cob wall

stone foundation laid in lime mortar

ground level

re-bar

rigid foam insulation

poured lime-crete foundation

The 'lime-crete' base should be wider than the foundation plinth and walls that it is going to support, so that the weight is evenly distributed. The depth of the trench into which the lime-crete is poured will be dependent on the local conditions of the site – drainage, subsoil, etc., and the size of building you are constructing. If in any doubt, consult a structural engineer to guide you through this important stage. To cut down on the amount of lime to be used, the trench can be first backfilled

If constructing a lime-crete strip foundation it is advisable to superinsulate this area so that the thermal efficiency of the whole building can be improved. This may be an occasion which calls for a material with a high-embodied energy such as a polystyrene foam board (see diagram above).

Its high-embodied energy will be counter-balanced by its efficiency and its ability to withstand moisture, as well as being reasonably priced and easy to get hold of.

Building with cob

Cob could be said to be the most democratic building material and process in the world. It is safe and accessible to all. For this reason women, children and the elderly can once again enter the realm of the building site.

Chapter 6

Building with cob is an extremely rewarding and fun process.
Make no bones about it, it is very physically demanding work, and
you will be sure to sleep well at night after a day of cobbing.
The best way to build with cob is with a team of people that you
really enjoy being with, and who are not afraid to get dirty.

The perception of cob building is that it is a slow process, because it is highly labour-intensive and because of the time needed to dry the cob in between layers (called 'lifts'). Because of this, it does not fit into today's mentality – that the most important element about creating a building is how quickly it can be put up. Despite being not entirely true (it is possible to start a cob building in the spring and have the roof on by the autumn), we prefer to see this as a positive aspect: it allows you to take a rest in between physically demanding building days, and hence creates a sustainable rhythm to see the building through to its end. It also dispels the modern-day notion that the building process is simply a means to an end, and awakens a new paradigm that the building process can be a highly creative and self-empowering means in itself. Whatever you choose to build, you will be creating a truly hand-made and original piece of work.

Tools needed for cobbing

The tools you will need for building with cob are refreshingly simple and low-tech. You can find most of what you need on your body, and after a quick rummage through your garden shed. The two most important tools are your very own hands and feet. Other than that, you will need the following items:

A sturdy garden fork for lifting cob up onto the wall

If you are of a lighter build don't be afraid to go for the smaller 'ladies' version – you will save valuable energy not having to lift a large hunk of wood or metal as well as the cob. Save it for your cobbing!

A variety of flat spades for trimming walls

The sharper the better, making it easier to splice the cob. These replicate the traditional paring irons used by cob builders of old. You can sharpen a spade end with a tool sharpener. Again, use size and weight according to your physical build.

tools

Some of your most important cobbing tools

Use a spade for digging and trimming, and forks for mixing, lifting cob and forming walls.

A mattock works well for shaping and an old saw is ideal for trimming cob when it is dry.

Use a level to ensure correct alignments and angles.

Wooden 'thwackers' to compact the walls from the sides when the cob is semi-dry. You can also use an old cricket bat.

Wooden mallet to compress cob in tight spaces.

Pick for excavating the clay subsoil from the ground.

Thwackers

A flat, paddle-shaped piece of wood to compress the cob from the side and blend the layers together. A cricket bat also works well.

Levels

Long, medium, small – to gauge the vertical levels of your wall. For walls that are tapering, you can duct-tape a piece of wood with your desired grade of taper to the level.

Wooden mallet

A woodworker's mallet works well to compress cob into tight spaces.

Skutch

For chasing out excess cob to fit lintels and create niches, bookshelves, and for any small trimming jobs.

Small mattock

For trimming walls from the side.

Buckets

You need at least a couple of sturdy buckets.

Tape measure

For making sure the wall width remains constant as you build up, and for measuring out positions of windows and openings.

Make your own thwacking tool

1 Take a log or round piece of wood – something that won't rot easily such as chestnut or oak. It should be about 750mm (2½ feet) long, depending on the size you are making. It is handy to make a variety of sizes.

2 Using either a chisel, chainsaw, hand-saw, sculpting axe, or whatever you feel comfortable carving with, first create a handle, which should be about two-thirds of the length of the whole thwacker. The diameter of the handle should fit your hand comfortably.

3 With the remaining round wood left at the top of the handle, shave a quarter of the wood off its face to create your flat 'thwacking' surface.

tools

The pitchfork method

1 Have your mix ready, prepared and sitting in a pile as close to the area that you are building as possible. If mixing by foot, it is generally best to spend the morning mixing cob and the afternoon building. Make sure the mortar in your foundation has gone off sufficiently to be able to support the weight of the cob.

2 With the pitchfork, take a scoop about the size of the head of the pitchfork and make into a small, square patty on the ground (see page 85). To make a patty: with your feet, create a square of compressed cob on the ground, making at least one fairly straight face which should be placed on the outside face of the wall. Think of these patties like creating bricks, which will make building and transporting easier. We don't encourage you to spend too much time on making the patties – a few whacks with your feet is sufficient.

3 Place the patty of cob on the wall and compress into place using consecutive blows with the fork head. You can also use your hands to position the patty on the wall for good fit. For a 600mm (2')-thick wall we find that three patties sitting side by side will make up the full width. Our standard practice for building a good wall and maintaining good internal and external faces is to first place one patty on the outside face, a patty on the inside face, and a patty filling the middle. All three of these patties need to be smashed together into one homogeneous mass. Continue this process along the wall. The amount that you build up in one session is called a 'lift' or 'perch'. Generally a lift should be from 300–500mm (12"–20") high, but will vary each time you build depending on how wet or dry your mix is, and the weather conditions. The stiffer the mix, the more it will hold its form and the higher you will be able to build without it bowing out ('splooging').

A warm, windy day will quickly draw moisture out of the cob, and it will dry much quicker than on a dull, damp day. Even though the former is more preferable, do be careful about too much direct sunlight, which will cause the surface of the cob to dry too quickly, leaving a crumbly, crusty outer layer. Generally speaking, the longer a cob wall can take to dry out, the harder it becomes. Traditional wisdom claims that the north wall is always the most solid because it has taken the longest to dry out. If your walls are in direct sunlight, protect with damp sheets or hessian sacks, hung in front of the wall. Never build when it is raining (unless you have erected a temporary shelter or have already built the roof). We have tried it, and it doesn't work – it makes for unhappy walls, unhappy people, and a sloppy end result. Go inside, drink hot tea, and gather your energy for when the sun comes out.

4 Repeat steps 1–3 until your first lift of about 300–600mm (1'–2') is achieved. Make sure that all the patties are blended

Building with cob

Scoop up forkful of freshly made cob close to your building site.

Make forkful of cob into quick patty with your feet for easier building and transport.

Scoop up patty with fork and place on wall. Compress into place using fork.

When a line of patties has been laid, walk on wall to ensure that they are thoroughly compacted into the wall below.

Trim off excess with a sharp spade when the cob is firm but not hard.

After trimming, stand on top of the wall and compress the cob from the side with a wooden thwacker

A pile of freshly mixed cob is best used within four weeks, although it will last indefinitely. Cover with tarps in between building to prevent it from drying out from the sun or getting too wet with the rain.

If it goes hard, simply add some more fresh water and re-stomp it back to a workable mix.

You don't need to spend too long making a perfect patty. This is just a helpful way to transport the cob to the wall, and makes it easier to build with.

The amount of cob you build up in one go is dependent on how wet your cob mix is and the drying conditions. A stiff mix on a warm, windy day will allow you to build more than a wet mix on a cold, damp day.

The amount you build in one session is called a 'lift' of cob.

On average, you can build a 300–500mm (12"–20") section in one session.

Top left: Placing the first lift.

Top centre: Forming the drip lip.

Top right: The finished drip lip in profile.

Left: Team-working on a larger walling project. Note the tarp keeping the cob malleable in dry weather.

Right: Trimming a second lift with a fork whilst the cob is still wet.

Above: Shaping the growing wall.

Above middle: Placing a high lift.

Far right: Trimming excess cob with a sharp spade once the cob is semi-dry.

Right: Trimming the freshly laid cob with a fork

together to become one monolithic mass. It helps to get up on the wall and walk along it compressing it into place with the weight of your body.

A note about the first lift

There are two things to say here: first, it is a vernacular practice to create a 'drip lip' to discourage rainwater from dripping into the stone foundation below. This involves building the wall 50–75mm (2"–3") proud of the stone foundation below it on the exterior of the wall. The end result is not only functional but is aesthetically very pleasing.

To create the lip (see page 86)

Allow the cob to splooge out 50–75mm (2"–3") over the stone foundation. This can be achieved simply by building the cob without trimming flush to the stone. Make sure however, that it is still a vertical line and is well compacted. Compact the bottom edge with your fork to a forty-five degree angle.

The second thing to say is: spend some time making sure the first lift is close to perfect, as it dictates where the rest of your wall is going. Make sure it is plumb on both sides and it is the right width. Try to avoid finishing a day's building halfway through a lift, and definitely avoid this when building across the top of a lintel.

5 As you go up in height, pay attention to the sides of the wall and roughly trim with your pitchfork to create a vertical plumb line. It also helps to compress the cob from the side using the fork head. You will trim the wall properly with a sharp spade once the cob has gone off a bit more.

6 Allow the cob to go off 'green hard' (i.e. it holds its form, but is still workable) for three to four days (depending on drying conditions) before beginning the trimming process. There are three ways we like to trim:

Spade trimming

Stand on top of the wall and grasp your spade with two hands. Use your body weight to thrust the spade downwards, slicing off any excess cob.

It is useful to have a tarp or board to collect the excess trimmings beneath the wall. These can be gathered, re-trodden if necessary, and re-used on the wall. If you have mixed your cob by foot, you will realise just what a precious commodity every piece of mud and straw is. If you are high up, take care not to thrust down too hard and topple forwards off the wall.

Mattock trimming (see page 82)

You can also trim standing in front of the wall – standing on the ground when the wall is low enough, or on scaffolding as you get higher – by using a small mattock to hack away the excess cob until level is achieved.

Old saw trimming (see page 82)

Using an old saw, stand to the side of the wall, using the straight edge of the saw to guide you, and with the saw pointed downwards, move along the wall cutting away at the excess cob. The flex in the saw is great for trimming curves and rounded corners. The old saw method is also good for doing an initial trimming of excess cob when it has just been laid and is fairly wet.

With all of the above methods, it is useful to use a long level and engage in a process of trimming a little, checking the level, and trimming some more until the right level is achieved. Always go slowly: it is much easier to take off more, than it is to put back chunks from over-zealous hacking. Also, don't over-trim as the cob will always shrink back slightly as it dries.

making contact with the wall with the flat side of the piece of wood. This really needs some weight behind it to achieve maximum compression. Again, take care not to topple forward off the wall, and hence don't make your thwacker too heavy to handle.

8 Repeat the above steps until the desired window and door height is achieved, and you are ready to put the lintels in place. A lintel serves to take the load of the wall and roof above an opening. The wood for an external lintel should be a hardwood such as oak or chestnut, so that it can withstand the elements over time.

> "Once we recognise that every situation is unique, and once builders begin not as mechanical executors of others' orders but as artistic individuals, even every door handle will be subtly different from each other."

Christopher Day, *Places of the Soul*

7 After you have successfully trimmed, you are ready to use a thwacker to compress the sides of the wall (see page 89). Thwacking has two purposes: firstly, to blend the different lift layers together; secondly, to compress the cob from the side. Be aware that thwacking will not straighten a splooging wall. You must always trim before you thwack. Stand on top of the wall, grasp your thwacker with two hands, and swing it back against the wall, each time

The wood for the internal and middle lintels can be of a softer wood, such as pine. The wood must be sturdy enough to take the weight of the cob and the roof that will be resting on top of it.

The lintels must span the full width of the wall. Two or three pieces of wood should be used to make up this width. For a dwelling do not use one large piece of wood, as this will create a potential cold-bridge from the cold outside air into the warm building.

The lintels must rest on wooden bearers which are placed into the wall, to provide a firm surface for them to rest and to disperse the weight above. Take a solid piece of 100x50mm (4"x2") hardwood, cut to the exact width of the wall. Chase away enough cob to fit this piece snugly into the wall so that the inside face of the piece of wood is flush with the inside face of the window or door reveal. The top of the bearer should be flush with the top of the wall and needs to be exactly level. One bearer needs to be fitted either side of the window or door reveal and these need to be exactly level with each other.

Once the cob is sufficiently hard under the bearers so that it can take the weight of the lintel without sinking, you are ready to put the lintel in place. To insert the lintels, rest either end on the wooden bearers, ensuring that the ends are resting with an overhang of at least 150mm (6") on either side of the wall. Once in place, check that the lintels are level (see pages 90–91).

9 Once you have reached the top of the first storey, and if you are going to create a second storey, you will need to insert floor joists to take the weight of the floor. It is important not to set the floor joists into a rigid mortar, such as cement, as they need to be able to accommodate the microscopic movements that will occur naturally in the building, especially during drying and settling. The best practice is to lay them into

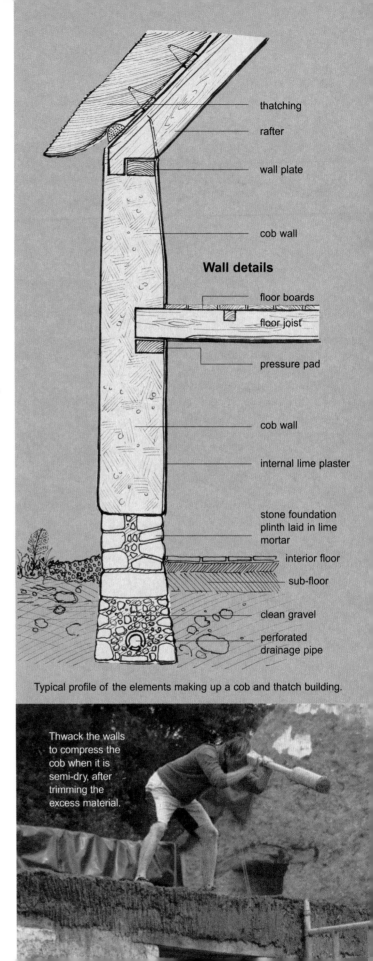

thatching
rafter
wall plate
cob wall

Wall details

floor boards
floor joist
pressure pad

cob wall

internal lime plaster

stone foundation plinth laid in lime mortar
interior floor
sub-floor
clean gravel
perforated drainage pipe

Typical profile of the elements making up a cob and thatch building.

Thwack the walls to compress the cob when it is semi-dry, after trimming the excess material.

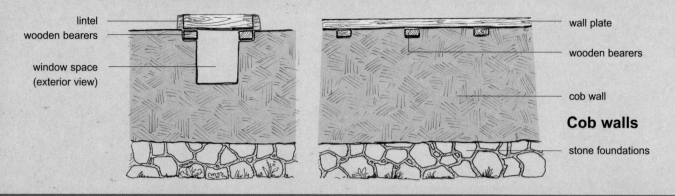

lintel
wooden bearers

window space
(exterior view)

wall plate

wooden bearers

cob wall

Cob walls

stone foundations

Top left: The lintel should sit on wooden bearers to evenly distribute the loads above the opening.
Top right: When you reach the top of the wall you need to introduce a wall plate to evenly distribute the load of the roof onto the cob walls. The wall plate sits on wooden bearers spaced roughly 1 metre (3' 3") apart.

the cob on wooden pads, larger than the joist end by at least 75mm (3") on both sides, so that the weight is evenly distributed on the wall. The joists must be inserted at least 200mm (8") into the wall.

The floor joists can be conveniently used as scaffolding, with planks laid across them for building the remainder of the wall.

10 Build up to the desired wall height and allow the cob to go off 'green hard' before putting in place the wall plate. The wall plate is a continuous, rigid timber plate that runs the whole perimeter of the wall, and serves to disperse the full weight of the roof across the width of the wall and around the building. It also provides a fixing point for the roof rafters and serves to tie the walls together.

Wooden bearers must first be put in place at one-metre (3' 3") increments along the top of the wall to provide attachment points for the wall plate. As described above, the wooden bearers must span the full width of the wall and must be sunk into the cob so that the top is flush with the top of the cob wall. The wood for the wall plate does not need to span the

full width of the wall, but needs to be at least 200mm (8") inches wide. A standard 8'x2' (2400x600mm) works well for this. It can be fitted in sections as long as the ends are butted tightly against each other over a bearer. Where the wall plate sits on the wall will be determined by the pitch and type of roof you will be fitting, such as thatch or slate. The wall plate must be attached to the wooden bearers using nails, screws or wooden pegs.

After a waiting period of roughly two months to allow the cob walls to dry and settle, your building will be ready to take the roof.

Tapering the walls

Cob walls can be tapered on the second storey of a building so that they are narrower at the top and wider at the bottom. This is possible because the top of the wall has less weight to bear than the bottom, which needs to carry the weight of the top storey and the roof. This has the benefit of reducing the amount of material needed, and therefore the labour needed to build it. There are no hard and fast rules about how much to taper, as long as the width of the top of the wall where the wall plate –

90

Left: Hardwood bearers should be placed into the cob either side of the lintel to disperse the loads above the opening.

Middle: a spirit level to ensure your lintels are level. Right: The levelled lintel in place.

and hence the roof – will sit, is no smaller than 450mm (18") wide.

To assist this process, work out your desired taper measurement, and cut a piece of timber on one side to the exact taper. Attach this to a long level with duct tape, and use it to create an even taper around the whole building on the internal face of the wall (see page 83).

the reveals of the opening are kept as level as possible as you build up otherwise it will make fitting your window/door frame difficult.

Alternatively, you can build a wooden former to the exact shape and dimensions of the desired window or door opening. Fit this in place as you build, and sandwich tightly with cob. Even though it is time-

 A lift should be from 300–600mm (1'–2') high, but will vary each time you build depending on how wet or dry the mix is, as well as the prevailing weather conditions. „

Creating windows and doors in cob walls

There are three options here: the most straightforward way to create window and door openings in cob walls is to simply leave a space for the opening as you build the wall up. It helps to use a single layer of cob blocks at the bottom to create a flush line from which to build up from. Ensure that

consuming to build the formwork, it has the benefit of creating perfectly formed openings and will save you time when you come to fit the windows and doors. Leave the formwork in place for at least three months (6 months if this is an arch form without a lintel) before removing. Lintels should be fitted on bearers above the formwork as you build, as described above.

Top: A completed upper lift.
Above: Ready to install the bearers and wall plate.

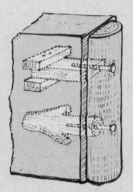

Wall anchors

Pieces of timber built into the wall will provide attachment points for window and door frames.

You can build a timber-framed window into the wall as you go along.

Windows can be cut out of the cob once it is adequately dried (6 months). This method allows you the flexibility to decide where you want your windows after the building is complete and saves time while you are building the walls up. However, it is a false economy, as it is an arduous and time-consuming task to cut out an opening through a 600mm (24")-thick wall. This should only be done for smaller windows.

You can also build in windows in their frames, or a frame without the glass, as you go. It is necessary to spike the frame with nails where it comes into contact with the cob to provide a good key. Expect a bit of shrinkage away from the frame, which will need to be filled in with cob or lime when the cob has fully dried out.

We do not recommend building in panes of frameless glass, as they are liable to crack as the building shifts during the drying process.

Fitting window and door frames

As you build up your cob wall it is necessary to build in pieces of timber into the sides of the window and door openings to provide attachment points for the frames. Often called wall anchors, these can be small pieces of 100x50mm (4"x2") or pieces of roundwood that sit flush to the face of the opening.

It is also a good idea to build in pieces of wood at other points in the wall to provide fixing points for cupboards, hooks and shelves, in fact any place where you know that you will need fastenings in the future.

92

Other methods of building with cob

The method described so far is just one amongst many ways to build with cob. Below are some others.

The hand method

If you really want to get down and dirty and engage with every piece of mud that you lay on your wall, you can do away with the pitch-fork and simply use your hands to transport and work the cob into the wall. Make sure that you continually walk on the wall to make sure that the cob is fully compressed. This is a great method if you are working on a small project and have the time to sculpt details into your wall. It is also good for getting into tight spaces, such as when filling under roof beams and other awkward areas.

JCB method

At the other end of the spectrum, some cob builders in Devon, such as Kevin McCabe, use the bucket of a JCB to lift large amounts of cob straight onto the wall top. This is then stomped in and compressed using foot and heel. The excess material or 'splooge' is pared off onto scaffolding that has been erected to help catch the trimmings. This is a great way to build up large areas of wall quickly.

Again, there is no right or wrong method. Which method you use will depend on your personal preference, the people you have available, the logistics of the site, and the size of your project. No doubt within one large project there will be a call for all three methods to be used at different times. Here are some points to be aware of when you first start building and the material and the methods are not familiar to you.

Some useful tips for first-time builders

Avoid shouldering

'Shouldering' is when the width of the wall diminishes as you build upwards, rather than remaining constant – which it should. Also, we cannot stress enough the importance of keeping the wall top level as you build up. If you do not maintain this you will find it hard to build up without the cob slumping off.

Three rules for avoiding shouldering, splooging and mushrooming (see above):

1 Constantly check that you are maintaining a level surface at the top of the wall so that you are always building on top of a flat, solid bed. This will prevent shouldering, and ensure a sound wall.

2 Constantly check the vertical levels of your wall as you go up in height, both inside and out.

Regularly use a level to check this, and trim wherever necessary. This will prevent splooging going unnoticed.

3 Establish what measurement your wall width is going to be from the beginning, and maintain this as you build higher. This will prevent mushrooming and will ensure plumb walls.

3 rules

Avoid splooging – caused by building up too quickly

We have described earlier that cob should be built up in 'lifts' that do not exceed the point at which the wall begins to splooge. One common mistake is to be too hasty in building up the wall, by either building up too much in one go or not allowing sufficient time between lifts for the cob to go off. If you end up in this situation, you will find that the freshly laid cob will slump and bulge. It may even cause large chunks to fall off the face of the wall. If this happens you need to slow down and allow the cob to dictate the pace of building. If you are building in damp weather and/or your cob mix is quite wet, this means that you will need to build up in smaller lifts. If slumping does occur, don't be disheartened – it is all part of the learning process and getting to know your boundaries with the material. If you are simply having problems with splooging material as you build, then take the situation in hand, slow down and give the wall some serious trimming. If a portion of your wall has slumped to the point of falling out, simply cut out a series of steps within the area of slumped cob to create a key for the new cob to be inserted. Allow the stepped cob to dry out a bit and then dampen down the surface before filling with new cob, ensuring the stepped area and the new cob are well bonded together. Dried cob blocks can also be used if the area is large. (See Chapter 12 on *Restoration* for instruction on how to use these.)

Avoid mushrooming

Avoid the tendency to build the wall wider as you go higher. This is one of the most common mistakes made by new builders, as it is so easy to do. Resist the temptation, and keep measuring the width of the wall.

Far left: Making up cob balls for the cob toss.

Left: Cob balls ready to be tossed.

Right: Tossing the cob balls along a line of workers.

Far right: A remote building growing from the ground without spoiling the surroundings by machine access.

If it goes unchecked for too long it will create pressure which bears down on the outer edges of the wall and may lead to an eventual collapse as you build higher.

Transporting cob from source onto the wall

The cob toss

The most fun you can have while building with cob is to be had in the cob toss! This is a great way to get the cob from the mixing site to the wall, get the heart rate up, bring people together as a team – and it's a fun excuse to chuck muck at other people!

Once your batch of cob is made, bend down, rip off a chunk of cob, and knead it into a round cob loaf. The loaves need to be fairly consistent and of a size compatible with your smallest cob toss team members.

Get everyone to make up a pile of loaves like making a stash of cannonball ammunition waiting to be fired. Create a line of people, with one next to or on the wall (depending on how high your wall is) to catch the loaves and place them on the wall. Then one person should be placed next to the cob loaves to start the toss, with people at intervals in between to catch and pass the loaves. Keep the pace steady enough so that the person on the wall has time to work each cob loaf in as it arrives at the wall. If you have a fast team, you can pile them on the wall and work them in when the cob toss is over.

Building when the wall gets higher – scaffolding options

For cobbing to remain comfortable and sustainable for the body, it is easiest when you don't have to lift your pitchfork full of cob above waist height. It is necessary to have all of your body weight to compact the cob.

This means that at some point in your build you will need to think about scaffolding. Below are some different methods to get the cob up onto the wall once it has gone above a comfortable building height. As always, there are different options depending on the size of the project and how much money you want to spend.

1 For a large project, you may feel justified in ringing the local scaffolding company and kitting the house up with professional scaffolding. If you decide to go with this option, it is necessary to leave enough room on the outside of the wall for trimming.

2 When you have reached the second storey and you have put in place your floor joists, you can lay planks of wood down on the joists. This will provide a stable and comfortable base from which to build your second storey. Trestles can be erected onto the planks when the walls get higher and when building up the gable ends.

3 A more flexible option is to use a moveable scaffolding tower, which can be manoeuvred around according to where you are building. These towers can also be heightened or shortened, according to the height at which you are building. Better yet, you can get scaffolding towers on wheels, which can be hired at most tool hire stores.

4 Other scaffolding options for lower heights include adjustable scaffolding, wooden trestles (which can easily be self-made), and home-made lean-to scaffolding made by 100x50mm (4"x2") pieces of wood. We had an American friend, Perry Uber, make us some of these while he was working with us on a project, and they successfully got us to the top of a two-storey building. These are much easier to move around than the towers.

How to get the cob up to the scaffolding at higher levels

1 For building at higher levels it will always be necessary to have more than one person on board. The simplest method requires someone to throw cob from the bottom up to a receiver standing on the wall or on the scaffolding. It works really well to have three people – one to make up patties, one to catch them on the scaffolding, and one to work them into wall. To throw cob up from the bottom requires some strength, so get your strongest team member in this position.

2 If you have gone for the professional scaffolding option, it will save you a lot of hard work to drop off large loads of cob with a tractor or JCB onto the scaffolding. This can then be forked onto the wall. Be careful not to load too much cob onto the scaffolding because it could cause it to buckle under the weight.

3 Yet another successful method we have used is a bucket and pulley system that can be attached onto the scaffolding railings. One person should be at the bottom, filling buckets from the pile and then sending the bucket up to the person on the scaffolding using the pulley rope. You can use a climbing pulley system (a Petzl), which can be

purchased at a camping store, or buy one from your local builders'-merchant.

Sculpting with cob

One of the great benefits of cob as a material is its flexibility and malleability. This allows you to sculpt in all sorts of unique features such as niches, bookshelves, benches and other furniture to your tailored specifications. A cob bench against a cob wall gives the impression that the furniture is an extension of the wall itself, as it gracefully follows the contours of your inside space.

To create a niche or bookshelf
It is easy to create niches or bookshelves after the wall has been built and had the chance to dry out for a few months. Simply use a skutch or other sharp tool to hack the cob out and sculpt your unique niche or bookshelf. Just remember, don't go more than 200mm (8") deep into the wall, as you may create a weak point. For larger arched niches, either use a wooden form (see section on doors and windows) or use a method from which we learnt at the Cob Cottage Company, involving a special mix of cob to corbel the arch into place.

To create cob corbels
1 Take a lump of cob and flatten with your palm.

2 Put six to eight strands of long straw in the middle and work these individually into the cob so that they disappear.

3 Fold the cob like a sandwich wrap so that the straw is sticking out at either end of the wrap.

Cob is very versatile. Even when dry you can retrospectively sculpt niches and bookshelves after the walls have been built. Left: Marking dimensions on the wall for the niche. Middle: Digging out the niche with a sharp spade. Right: The finished niche used for flowers, candles and other special treasures.

4 Fold in the end bits so that the straw pieces sticking out from the ends disappear.

The straw acts as extra reinforcement to strengthen the arch or cantilever and avoid collapse.

To build using the cob corbels

- The cob corbels are gradually cantilevered from each wall so that they eventually meet in the middle to create an arch. They can also be used for relief work.
- Each successive corbel must be worked in so that the cob becomes one homogeneous mass.
- The arch must be built out gradually to avoid collapse, which means only building a few corbels on each side at one time, then allowing these to dry before applying the next batch.

Use these to make arches for niches, bookshelves, windows, doors, gateways or arched fireplaces.

Alternatively, you can create a square niche by using a lintel in exactly the same way you would when creating a window or door opening.

To build a bench

1 A bench can be built directly onto the ground in an inside space, but a small foundation of either stone or brick will reduce damage if a flood were to occur. Laying a foundation will also help to define the shape and size of the bench when you come to lay your cob. If building an outside bench, the foundations must be at least 300mm (12") off the ground, and the bench must have a small roof over it to protect the cob from the rain.

A cob arch constructed out of 'corbel cobs', by *The Cob Cottage Company*, Oregon, USA.

A cob bench is easy to construct at the time of wall raising, and will appear to grow from the pleasing curves of the walls. Cob bench by *Seven Generations Natural Builders*, Oregon, USA.

2 The cob mix used is exactly the same as that which you would use to build your cob walls. Simply build as described above up to your desired height, sculpting the bench to your personal design. Test your bench out so that it feels comfortable to sit on.

3 Once your bench is complete, you can simply plaster it in to match the rest of your interior walls.

Plumbing & electrics

Plan ahead of time exactly where you want to place your electrical and plumbing services. We advise that you consult a professional plumber and electrician about the layout you have chosen before building commences. Cob is very safe with electric cable against it, but wires should still be placed inside conduits. Leaking pipes could be damaging, so make sure your arrangement and fitting are sound.

Electric cables can be fitted into the walls once they have been built simply by chasing away channels in the cob before plastering. Use large 'U' nails to tack the cable in place into the wall. The light switch box can be nailed into place as in a conventional house.

Use screws with a grooved plastic sheathing for attachment. Fill the chased-out channels with built-up layers of cob or lime plaster to encase the cable. Alternatively, the wires and box can be incorporated into the walls as they are built up. To anchor the box deep into the centre of the wall, attach it to a U-shaped anchor using timber scraps and screws. They can also be fitted in the floors between the floor joists, and in interior partition walls and ceilings. Plumbing pipes can be laid in much the same way as in a conventional house. They are best laid between floor joists, in ceilings, and interior partition walls, but can be set into a cob wall. It is recommended to take some extra precautions to avoid damage in the cob walls in the event of a leaking pipe. Ensure that there are no joints in the pipes wherever they are buried into a cob wall, and encase the pipes in larger plastic tubing so that if there is a leak the water will be caught in the tubing and will not damage the wall.

If you choose to lay an earthen floor, leave a movable, wooden strip along the edge of the wall under which you can bury your water pipes and conduits. This will enable access if needed at a later date.

How cob dries

Cob dries from the outside in, and in a British climate it can take up to a full year for the cob to fully dry and to complete its shifting and settling process. The most important element to cob drying is good air circulation to whisk away molecules of moisture out of the body of the cob wall.

Obviously warmth helps this drying process, but cold air circulation is just as effective. If you are building in an enclosed space it is a good idea to keep the windows and doors open as much as possible to ensure a good circulation of air.

Damp, windless days are the worst conditions for cob drying, because the moisture has nowhere to go other than stay in the wall. This is why cob building has a definite season. The old saying goes, as cob revivalist Alfred Howard states: "Cob building should start when the birds begin to build their nests, and the roof should be on when they begin to lose their feathers."

In a British climate, the ideal time to start building is mid-April, when the days have started to warm up, and finish in mid-October to avoid the cold rain. We have built throughout the winter months but found progress to be slow and frustrating. We now

protected from heavy downpours and frosts throughout the winter.

Cob garden walls & courtyard spaces

Cob was traditionally used for garden and boundary walls, and all over the United Kingdom examples of such walls still standing, and serving their original function, can be found. In the garden, walls made out of cob can utilise the ability of the cob to absorb heat from the sun in the day, which will be released at night when the temperatures drop.

On a sunny, south-facing wall, this can create a microclimate suitable for growing

> "Cob walls for garden fruit are incomparable. They retain the warmth of the sun and give it out through the night, and when protected on top by slates, tiles or thatch will last for centuries.
>
> Sabine Baring-Gould

choose to honour the seasons, working with the material instead of against it, and look forward to a rest in the coldest winter months.

This will also encourage you to stick to a good building schedule with a definitive deadline to get the building roofed by the end of autumn. It is very hard work and very stressful trying to keep cob walls

exotic fruits. It can also make for a marvellous place to sit and enjoy the warm rays of the sun late into the evening and night. To quote Sabine Baring-Gould in the late nineteenth century: "Cob walls for garden fruit are incomparable. They retain the warmth of the sun and give it out through the night, and when protected on top by slates, tiles or thatch will last for centuries."

All pictures this page: A walled courtyard under construction at Mount Pleasant Eco-Park, Cornwall. Constructed through a training project led by *Cob in Cornwall.*

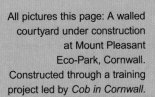

Left: Note the hardwood bearers underneath the lintel to support the loads above the opening.

Above, left and right: Supplies are handed up by fork to create a 'cob triangle' to form the pitch for the roof as an alternative to using wood. Note the bearers embedded into the cob for the wall plate to attach to.
Right: The cob triangle is capped with scantle slates embedded into a lime mortar. Note the large overhang to protect the walls from the rain.
Scantle slates by Ben Verry.

During building, you need to protect your cob from the elements – rain, sun and snow. Tarpaulins are also useful to control the drying rate, especially during a hot spell.

Constructing a cob garden wall is a great place to start experimenting with cob and a great way to get to know the material.

Protecting walls during building

It is essential to protect all cob walls from rain while building, because heavy downpours and steady rain will certainly sabotage a freshly laid wall. At the end of every working day, make it a habit to cover the walls, no matter how settled the weather looks. British weather is a temperamental beast, and we have been caught out many a time! Standard tarps anchored down by rocks and timber work well if you are regularly removing the tarps during the day in between building. If the tarps are not regularly removed, condensation will collect under the plastic and this will slow the drying rate of the wall. To prevent this condensation from affecting the wall, put a layer of straw or hessian sacking between the tarp and your new lift.

" ... then they procure from a pit contiguous, as much clay or brick earth as is sufficient to form the walls, and having provided a quantity of straw or other litter to mix with the clay, upon a day appointed, the whole neighbourhood, male and female, . . . assemble, each with a dung fork, a spade or some such instrument. Some fall to working the clay or mud by mixing it with straw; others carry the materials; and four or six of the most experienced hands build, and take care of the walls. In this manner the walls of the house are finished in a few hours after which they retire to a good dinner and plenty of drink which is provided for them, where they have music and a dance, with which they conclude the evening. This they call a daubing. "

Sinclair, *Terra Britannica* (1792)

Roofs

7

Traditionally, the construction of a roof, and the materials
it was made out of, were directly related to the immediate
natural surroundings. This ensured that the roofing
system was suited to the local climate, and blended
seamlessly into the natural landscape.

Chapter 7

A ny roofing system has many functions, and it is an extremely important element to get right. The moment the roof materialises on a new structure is the moment it is transformed into a building, and can fulfil its requirements as a shelter and a home. When designing a roof and deciding what materials to use, consider the following points. A roof should:

1 Protect the building and its inhabitants from the elements of the wind, rain, sun, and snow. Unprotected walls or a poorly fitted roof will cause water to penetrate into the fabric and lead to structural deterioration. This is especially important when constructing walls out of cob, which are vulnerable to water damage.

2 Insulate the building from unwanted losses and gains of heat and cold.

3 Shed and direct rainwater and snow away from the walls.

4 Provide aesthetic consistency within the context of the whole building. For example, a Spanish clay-tiled roof may look awkward resting on a cob cottage.

Opposite: Timber frame at Chithurst Buddhist monastery, designed by EcoArc Architecture, built by Green Oak Carpentry Co.
Below left: Roof structure made from unmilled roundwood collected from the surrounding woods.
Below right: Traditional cruck timber frame being lowered into place by workers from Carpenter Oak Ltd.

In addition, a roof can provide you with:

1 An opportunity to collect rainwater for use around the home and garden.

2 A high vantage point on which to situate photovoltaic systems for energy production.

3 An opportunity to provide green living spaces with turf roofs – especially desirable in urban environments.

The roof system itself comprises two main components:

1 The structure, which gives the roof its shape and strength (i.e. the timber frame).

2 The protective skin, which is attached to the structure and provides waterproofing and protection from the elements (e.g. thatch or slate).

In modern, conventional buildings it is commonplace to install a synthetic waterproof membrane between the structure and the outer skin. In roofs made out of natural materials, this is not always necessary, and indeed much of the time their very success and health relies on their ability to breathe naturally, which is dependent on the free movement of air.

As with all vernacular building practices, traditionally the construction of the roof and the materials it was made out of were directly connected to the immediate natural surroundings. This ensured that the roofing system was suited to the local climate, and blended seamlessly into the natural landscape. For example, the steep-pitched timber roofs of northern Europe were constructed from the abundant pine forests, and angled to prevent snow from building up and overloading them. In the desert, where there are no trees, vaulted roofs are made out of earth blocks, designed to maximise cool airflow around the building to relieve the heat of the day.

Traditionally, most cob buildings in the UK were covered with a roof of thatch attached to a timber frame. This was because most cob cottages were constructed and lived in by the rural workforce, who had at hand the waste straw or wheat reed that remained after the harvesting of the grain.

A traditional timber frame is the perfect accompaniment to a cob building. The chunky timbers and traditional methods of construction will complement the thick, rounded cob walls.

These methods can be executed in a way that brings a fresh, contemporary feel to the overall building. The UK has a rich history of such methods, and many contemporary timber frame companies and individuals who are producing a very high level of craftsmanship and quality of design.

A new cob building does not need a special roof design. However, special consideration must be given to the areas of the overhang

Timbers – milled, sawn, hewn or left in roundwood form – all have their own charm and find application in buildings from barns to cottages.

Top left and right, middle left and centre: Timber frame constructions by Carpenter Oak Ltd.

Above, right and far right: Roundwood timbers sourced from a nearby forest complement the rustic charm of this rural studio designed and built by Cob in Cornwall.

Left: Thatched granary, Africa. Right: The round curves on a roof, such as on this Victorian garden folly replica, fit beautifully into the wood it is part of. Alice's Seat, Trebah Gardens, Cornwall. Thatch by Mike Pawluk.

of the eaves, and the distribution of the weight of the roof on the walls. In thatch or turf roofs with no guttering, the overhang must be at least 450mm (18"), to ensure that rainwater has adequate clearance from the walls as it falls to the ground.

As discussed in Chapter 6: *Building with cob*, a wooden wall plate must be inserted onto the top of the cob wall to ensure even distribution of the weight of the roof. The wall plate also acts as an attachment point for the roof structure, and anchors it securely to the walls below.

The freedom of design that cob allows offers the potential for some unique roof designs. However, unless you are fully experienced in roof design and construction, always have your ideas approved by a professional engineer, timber framer or specialist builder. You cannot afford to make any mistakes in this area.

It is obviously up to the preference of the individual and the materials available in the locale, to determine what roofing system is chosen, but there are four options, all made from natural materials, which we consider best suited to a cob building. These are:

- Thatch
- Wooden shakes and shingles
- Natural British slate
- Turf

Left: A traditional cob cottage with a well-maintained thatch roof.
Right: The golden tones of a newly thatched building.

Thatch

A thatched cob building brings together the four most simple, readily available natural building resources on the planet – grass, mud, stone and wood. From the cob cottages of rural Cornwall to the mud dwellings of Burkina Faso in Africa, these four materials have been brought together since the beginning of shelter building to create structures of unrivalled comfort and beauty.

Thatching – broadly defined as the use of any vegetable material as a roof covering – is the oldest and most common roofing technique in the world. In the UK it was applied to nearly all smaller houses until the

end of the Middle Ages, and until the mid-19th century in rural areas. As mentioned earlier, the most common material used – straw or wheat reed – was generally a by-product of the wheat that was grown and harvested for grain. It was very cheap (if not free) and demonstrates the ingenuity of our ancestors: how they efficiently utilised local resources, out of practical necessity. The decline of thatching as a roofing material occurred as a result of several factors.

Changes in farming methods, and the mechanisation of the methods that were used to thresh the wheat – from the simple hand sickles and reaping hooks of the Middle Ages to the modern combine harvester – meant that the straw was bruised and damaged during the processing, which led to a decline in the quality and durability of the thatching material.

The emergence of cheap transport and building of canals meant that the movement of other roofing materials around the country, especially Welsh slate, became commonplace and provided financially viable alternative options to thatch as a roofing material.

Natural insulation

Thatch is an excellent natural insulator. The hollow stems of the straw efficiently trap air, thus preventing any unnecessary loss or gain of heat and cold. A 400mm (16") thick covering of thatch has an excellent U-value rating of 0.20W/m²K, which is half that of an uninsulated slate or tile roof. Thatch is also similar to cob in that it is a great

The thatched roof is both a masterpiece of technical ability and a feat of artistic skill, and the rural thatcher is a craftsman in the truest sense of the word, embodying those personal skills and values which are becoming increasingly rare in a world in which standardisation and mechanisation are the accepted norms both in life and in work.

Barry O'Reilly, *Living Under Thatch*

As house insurance gradually became more and more common, especially when tied in with mortgages, thatch roofs became extremely expensive to insure, owing to the fire risks involved.

Despite the above points, thatching is beginning to regain popularity; not just because it is aesthetically pleasing, but because it has many functional and practical advantages. As a 100% natural roofing system, it scores highly with regard to its environmental profile. Thatch roofs are beneficial because they include the following:

regulator of temperature, which allows for the maintenance of an even temperature throughout the year, helping to keep the internal space cool in summer and warm in winter. Thus, thatch is an extremely energy-efficient roofing system.

When installing a thatch roof on a cob building, it is necessary to add an extra layer of insulation, such as sheep's wool, to compensate for the high U-value of the cob walls. This will improve the overall thermal efficiency of the building. For more information on this, refer to Chapter 8:

Insulation, and Chapter 13: *Planning permission and building regulations.*

Renewable Resource
The reed and straw used to make the thatch is a 100% renewable energy resource, which can be planted year after year and therefore can be used indefinitely. Additionally, thatch made from wheat straw is the by-product of the grain threshing process, which in itself is a relatively non-polluting activity, requiring little energy or sophisticated tools. This is especially true when grown and harvested with more traditional methods. It also has the potential to be a resource produced on a completely local basis.

100% biodegradable
The rotten straw taken from a roof that is being repaired or re-thatched can be recycled to make nourishing compost for the garden. Traditionally, especially in Scotland and Ireland, roofs that were soot-laden from interior open fires were completely stripped every two to three years, composted on dung heaps or spread directly onto the fields and gardens as a rich manure (see David Iredale and John Barrett, *Discovering Your Old House*). Compare this with a roof which is made of asbestos or cement slates.

Aesthetically pleasing
A simple thatch roof, especially when situated in a rural setting, will visually blend into the surrounding fields, especially when they are filled with golden wheat. Like cob and wood, a thatch roof will age gracefully, changing from a golden colour into a more mellow shade of silver-brown.

"Note too the change of colouring in the work as time goes on; the rich sunset tint, beautiful as the locks of Ceres, when the work is just completed; the warm brown of the succeeding years; the emerald green, the symptom of advancing age, when lichens and moss have begun to gather thick upon it; and last scene of all, which ends its quiet, uneventful history, when wind and rain have done their work upon it, the rounded meandering ridges, and the sinuous deep cut furrows, which, like the waters of a troubled sea, ruffle its once smooth surface." (P. H. Ditchfield, *The Charm of the English Village*).

The challenges of thatch
As with all things, there are some less than positive aspects about becoming the proud owner of a new thatch roof. These include:

High labour costs
Thatching is a highly skilled craft, and requires many years training and practice. In this respect, it is not for the self-builder. Additionally, it is extremely labour-intensive, making it a relatively expensive option. It is ironic that – as is the case with most traditional skills practised in the modern world – thatching was at one time predominantly the roofing method of the poor rural labourer. However, if your design and building are simple and of modest size, we feel that it is a method worthy of its extra cost, especially due to the fact that most

contemporary thatchers really love their work and take pride in their craft. This can really be seen and felt in the final outcome.

Fire risk

There is an assumption that thatched roofs have a greatly increased risk of catching fire. However, English Heritage, in their booklet with the breathability of the thatch, and means that it cannot be appreciated in its natural state from underneath. If fitting a fire-retardant membrane, it is essential that there be a significant air gap between the inner face of the thatch, so as to provide adequate ventilation for the thatch layer.

> It is charming in its youth, maturity, and decay.

P. H. Ditchfield, *The Charm of the English Village*

'Thatch and Thatching, a guidance note' (2000) claim that the insurance records argue against this. By taking certain precautions, this risk can be greatly diminished. Current building regulations stipulate that all new thatch should have a fire-retardant membrane between the thatch and the ceiling. In one respect, this is a shame because it interferes

Additionally, there are many precautions that can be taken in regard to chimneys, electrical wiring (minimise the amount in the ceiling) and general household activities, which will greatly diminish the possibility of a fire. In areas where thatching is common, a local thatcher or the local fire department will be able to advise you about these.

Nothing surpasses the smell, look and feel of a newly-thatched building. Thatch may last up to 40 years, and once it has surpassed its natural lifespan will find use as compost for field or garden.

Different types of thatch and their lifespan

The lifespan of a thatched roof is dependent upon the type and quality of the thatching material used, how well and how thickly it is laid (400mm/16" is the suggested thickness), how well it is maintained, and the local weather patterns. In the UK, all types of grass material, including heather, were traditionally used, but today three main types of cereal grass are used:

- Long straw (uncombed wheat reed)
- Combed wheat reed
- Water reed

Water reed has the longest life expectancy, at fifty to eighty years; then combed wheat reed at thirty to forty years; and finally long straw at ten to fifteen years. The latter two are grown abundantly throughout the UK, but as mentioned earlier, modern harvesting methods and the use of nitrogen fertilisers that weaken the grass has rendered them virtually unusable for thatching. To make this an option, the wheat is best grown organically and harvested in the traditional way to produce a strong material. Water reed, used mainly in East Anglia and Scotland, needs to be grown in marshy or wet conditions.

Its production in the UK has greatly diminished because of reduced demand and the drainage of much natural marshland. This means that unfortunately the best material has to be imported from Europe. The best water reed available in the UK is Norfolk Reed from East Anglia. A compromise in regard to the principle of buying local may need to be made, as the imported reed is of higher quality and will therefore last longer.

Maintenance

Thatch roofs will reach or exceed their maximum life expectancy if they are well maintained. This includes repairing patches pecked by birds or damaged by winter storms as soon as they appear, and keeping plant growth overhanging the roof to a minimum.It is best to avoid a covering of wire meshing over the thatch as is regularly seen (to deter birds and vermin) because this hampers the natural water shedding principles of thatch, which work as follows: the water runs unrestricted from straw to straw down the steep pitch of the roof (minimum 45°, ideal 50°), and falls off the eaves before having the chance to penetrate into the thatch. If wire mesh is used, the rainwater is deflected across the mesh surface and is therefore more likely to soak into the thatch, hence reducing its lifespan.

Life expectancy can also be affected by the geography of the building: in the wetter western counties, it will obviously be lower than in the drier, colder eastern counties. On a site-specific level, orientation is important, with a potentially increased life expectancy given to buildings that have a gable end facing the prevailing wind and rain, as opposed to the whole side of the building. In a passive solar design, however, this may be different from the best orientation for solar gains, and will need to be considered within the whole scheme.

Wooden shingles or shakes

Wooden shakes or shingles (shakes are split, shingles are sawn) have been used since the Bronze Age as an effective roofing material. In the UK, oak shingles were used in the Middle Ages as an alternative to thatch. Some can still be seen today on church spires around the country, especially in the south-east of England.

For example, Salisbury Cathedral was originally roofed with shingles from the New Forest. Their use diminished mainly due to the arrival of new roofing materials such as clay tiles, from the 16th century onwards.

Today, they are gaining in popularity again as a roofing system with good environmental credentials, as long as they are purchased from a certified sustainable source and are not fireproofed and preserved using carcinogenic substances. They are from a potentially renewable source, and are 100% biodegradable. There is little energy needed in their manufacturing. They have a relatively high insulation value, and do not require a waterproof membrane underneath them, as they need to be able to breathe and dry out between rainstorms. Wood shingles or shakes should not present a major fire hazard in the UK, owing to our wet climate. They do require a bit of extra maintenance to ensure that they have a long life, mainly in regard to preventing the build-up of dirt between the shingles, which may cause rot if allowed to remain.

Aesthetically, a shingle or shake roof complements a cob building owing to its potentially irregular look (if random-width shakes/shingles are used) and because it can bend around curves – great for circular or contoured cob structures. Shakes and shingles also have the advantage of being extremely light, and therefore require a less sturdy roof structure than is needed for the heavy loads of thatch and turf roofs. As with all the other natural roofing materials, they age well, softening and mellowing in colour over time, and therefore will blend quietly into the surrounding environment.

The UK has a number of sources for sustainably produced wood shingles and shakes, which are listed in the *Resources* section at the back of the book.

Natural slate roofs

Natural slate roofs became the common roofing material throughout the UK from the mid-19th century on, when canal transport enabled them to be distributed economically, mostly from Wales and Westmoreland. They quickly superseded thatch. There are three main sources of naturally quarried slate in the UK: North Wales, Cornwall and the Lake District. Historically, these were all locally based industries. Today, slates are sourced from all across the globe: Europe,

Various roofing materials which complement cob. Top left: Clay roof tiles. Top middle: New cedar shingles. Top right: Earth-block dome roof in Iran (earth roofs are only suitable for climates with very little rain!).

Bottom left: Natural slate. Bottom middle: Corrugated metal roofs are commonplace on old cob agricultural buildings, which have been traditionally put up as practical replacements for worn-out thatch roofs. They look attractive next to the red and golden tones of the cob. Bottom right: A roof made out of long, overlapping cedar shakes.

South America, Asia and Africa. They are sadly much cheaper than the UK-based industries because they are produced with much cheaper labour costs. It does not seem right to be sourcing slates from halfway across the world when we have our own quarries in the UK, even though it saves money. The wide variety of hues and tones that can be found in British slate, from the greens and purples of North Wales to the greens and reds of Cornwall and the greens and blues of the Lake District, all complement the earthy colours of a cob building.

A slate roof needs to be insulated with at least 300mm (12") of insulation to bring it to a U-value of 0.13. It is beneficial to go above the standards required, reaching a thickness of up to 450mm (18").

Above: This cob building, with a roof made from turf excavated from the building site, gives the impression that the building has literally been raised from the ground. Building designed and built by Cob in Cornwall.

Turf roofs

Traditionally, turf roofs have largely been produced locally by people building for themselves. The methods are simple, and the grass and earth resources are infinite and can be used direct from their source.

Turf roofs are currently gaining in popularity within the green building movement as a sustainable roofing system. Contemporary turf roofs have been inspired from the ancient sod roofs of northern and central Europe, especially throughout Scandinavia and Switzerland. Like thatch, turf roofs utilise grass as a roof covering, but differ in that it is a living system, which can support a wide range of plants, birds, and insects.

A turf roof consists of a covering of vegetative matter, which grows out of a layer of soil or organic matter on top of a waterproof membrane. At their simplest, they consist of a covering of grass, but can be created to support anything from herbs, vegetable and wild flowers, to even bushes and small trees.

Why build a turf roof?

Turf roofs are beneficial to the environment and the people who live under and around them. They are an excellent roofing system for self-builders because they are relatively simple and inexpensive to construct. For natural builders, they can fulfil a desire to use predominantly local, naturally sourced and recycled material.

Efficient utilisation of materials

Turf roofs can be created to become a literal extension of the ground below. You can use the topsoil and turf that have been removed from the building site during clearing and preparation. During excavation, separate out the topsoil, carefully store the turf in a shaded area, and water regularly until used. This provides a very satisfying utilisation of 'waste' excavation material; it also ensures that no habitats are lost and that the building will genuinely be a part of the environment.

Aesthetic sensibilities

Natural living roofs are a perfect partner to a cob building, and again work off the magic synergy of the four basic natural resources –

Middle: These Icelandic turf-roofed dwellings blend effortlessly into the mountain peaks of the surrounding countryside while giving excellent insulation. Right: Roundwood or milled timbers can be used for the fascia board.

stone, earth, wood, and grass. Turf roofs will complement the soft, rounded nature of cob walls and add to the natural essence of the overall building.

Excellent thermal insulation

Turf roofs are excellent at providing thermal insulation because of the thick blanket of earth, and because of the ability of the plants to trap pockets of insulating air between them. In the winter they help to keep the building warm, and in the summer to keep it cool by absorbing heat from the sun. This can be beneficial in urban areas, where the excess of concrete buildings can create a 'heat island' effect, causing a rise in temperature of up to 5°C. This aspect is especially beneficial as the effects of global warming begin to be felt.

Rainwater absorption

Turf roofs have the ability to absorb up to 85% of storm rain, so the amount of water entering the drains is much reduced.

Improved air quality

Turf roofs can improve local air quality, which is especially pertinent in an urban environment. The plants can absorb carbon dioxide in the atmosphere from the excess levels of pollution in the environment, and release oxygen.

The provision of animal habitats

In an urban environment, turf will provide habitats for insects, which will encourage the proliferation of insect-feeding birds.

Perfect for the self-builder

Unlike thatching, which is a specialist craft, this is a fairly straightforward roofing system for the self-builder, which can be constructed using simple hand tools, and requires little previous experience. It can generally be created out of materials naturally sourced from the site or the local vicinity.

Long lifespan

Turf roofs can last indefinitely if they are well maintained, well constructed, and a good quality waterproof membrane is used (an EPDM liner is best – see next page).

The provision of green space

In urban environments, turf roofs can provide attractive and much needed green spaces and areas to grow food.

Layers of a basic turf roof

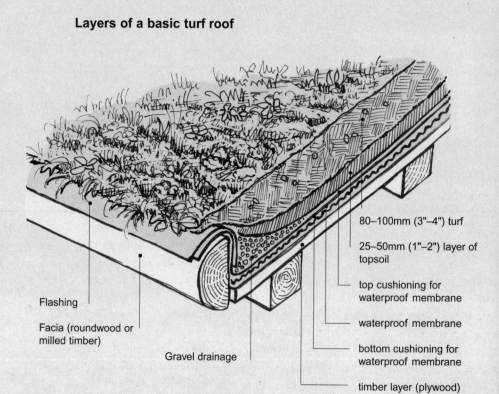

80–100mm (3"–4") turf

25–50mm (1"–2") layer of topsoil

top cushioning for waterproof membrane

waterproof membrane

bottom cushioning for waterproof membrane

timber layer (plywood)

Flashing

Facia (roundwood or milled timber)

Gravel drainage

How to lay a turf roof

Things to consider when laying a turf roof:

1 Turf roofs are very heavy (two to three times heavier than a slate roof), especially when wet or covered in snow. This means that it is essential that the roof structure supporting it be built to accommodate this. Turf roofs spanning a large building will need to be accommodated by carefully considering the design of the whole structure so it can take this extra weight.

2 Turf roofs require the use of a synthetic rubber liner; these use a very high amount of embodied energy to produce, as well as being expensive. Traditionally, sixteen layers of birch bark were used in Scandinavia until the mid-20th century, and although this is not a viable option in 21st-century Britain, there are currently some other eco-friendly options on the market.

However, EPDM liners (ethylene propylene diene monomer) cause minimum environmental damage as they are extracted from monomers. Additionally, they are extremely durable, effective, and will last indefinitely if protected from the sun.

They can also be re-used if not glued or nailed in place.

3 If a leak does manifest itself in a finished roof, it can be difficult to locate the source to carry out a repair. However, this can be avoided by observing some simple precautions:

- Never walk on the membrane in shoes during the construction process
- Obsessively check the roof structure for any protruding nails, staples, stones, sharp slivers of wood, and any other sharp objects that may puncture the liner.
- Cushion the liner from below and above with carpet underlay, matting, cardboard, or any other soft material that will help to protect it. (See laying steps below)

The roof structure for a turf roof needs to be at a pitch of 25° or less (it can be flat), to ensure that the turf doesn't spend its life trying to defy gravity. If a turf roof is laid onto too steep a pitch and the turf slips to expose the waterproof membrane beneath, premature disintegration of the membrane can occur from exposure to the UV radiation of the sun.

Again, ensure that the roof structure has been designed and built to be able to take the weight of the turf – consider also the extra weight of the already heavy turf when sodden after a rainstorm or laden with snow.

Layer 1 – Base timber layer
On top of the roof rafters, lay a solid timber base from which the turf roof covering will be supported. This can be plywood sheets, floorboards, or other solid timber sheathing.

At the lowest points of the roof, it is necessary to insert a drainage release point to drain excess water off the roof. A standard shower tray drainage piece, fitted snugly into a hole created in the timber sheathing, works well. Make sure your liner is tucked securely round the drainage tray.

Fascia board
It is essential to attach a fascia or edge board around the whole perimeter of the roof. This serves to contain the turf so that it doesn't slip down off the roof, and also provides a lip for the waterproof membrane to be draped over. This can be constructed out of any durable wood – roundwood, bendable plywood, or thick milled wood. The design is flexible as long as there is at least a 150mm (6") lip above the roofing base (Layer 1).

Layer 2 – Cushioning for the waterproof membrane
This consists of covering the whole of the solid timber base layer with a soft cushioning to prevent the waterproof membrane from being punctured by any sharp objects. Soft carpet underlay works very well, and can be easily laid. Try to find someone who has just taken up an old carpet, or ask for scraps at a carpet store.

Layer 3 – Waterproof membrane
This is the most essential part of the turf roof. For a long-lasting watertight roof, it is worth

investing money in an EPDM membrane (see *Resources* list for suppliers). Its longevity and effectiveness makes it a worthwhile sacrifice for your pocket and for the environment.

Make sure you observe the following:

- Ensure that you purchase an amount of membrane that exceeds the total roof area, so that you can cut it down to size, and so that the edge will overlap the fascia – it can be cut back to fit once the turf is in place. It is also important to leave some give in the membrane as you lay it, especially if laying in warm weather, as the membrane will contract during cold periods.
- Do not wear shoes while laying the membrane, and avoid walking on it at all once it is laid. This is the most common time when the membrane is damaged, which will cost you in the future.
- It is best to use a whole swathe of the same piece of material to avoid vulnerable seams.
- Take care in fitting the membrane around openings in the roof, such as around chimneys and flue pipes: ensure that the membrane fits very tightly, and insert the appropriate flashing material.
- Create the appropriate holes in the membrane to line up with the drainage pieces.

Layer 4 – Second cushioning layer

The function of this layer is to protect the membrane from damage from the soil, and as a rooting medium for the vegetation above. This can comprise the same material as in Layer 2. Take care when laying this over the waterproof membrane, and take the precautions mentioned earlier.

Layer 5 – Gravel drainage

A small layer of gravel 150mm (6") wide, needs to be laid around the perimeter of the roof, up against the fascia board. This is to prevent the accumulation of soil in the drainage piece that would block the free exiting of water. It will also help to prevent water from pooling on the roof.

Layer 6 – Layer of organic matter

To help the turf to get established, it helps to lay a layer of topsoil on top of the cushioned layer, 25–50mm (1"–2") thick.

Layer 7 – Turf layer

For an instant green roof, locally cut turfs (best from the land surrounding the building, or saved from the excavations from the ground works) can be used. These can be cut in 300mm (1') squares, and need to be 75–100mm (3"–4") deep. Alternatively, a local turf company may have second grade turfs, which aren't good enough for lawn laying, which they will sell at a discount rate. Simply arrange them on the roof so that they are butted tightly up against each other. The best time to lay the turf is late autumn, so that they can benefit from the ensuing rainy season. Turf laid in the summer or in a dry spell of weather will need to be well watered so that the turf does not die.

An alternative to laying cut turf is to seed the compost/topsoil layer with grasses or meadow flowers of your choice. If you choose this option, the compost/topsoil layer will need to be at least 100mm (4") deep. Although this option will give you more flexibility in what vegetation you have on your green roof, it makes the soil more vulnerable to slippage in the early days before the roof system gets established. A more expensive option is to buy ready-to-lay sedum.

Optional insulation layer

For extra insulation, cork or foamed glass can be installed on top of layer 1 (the base timber layer). The material used must be able to withstand contact with moisture.

Maintenance

A turf roof requires very little maintenance, especially in the UK where long periods without rain are rare. If in the summer the roof begins to look very dry, a simple watering will revive any flagging grass and encourage new growth. It is, however, possible to allow the grass to die back and wait for rain to regenerate the roof with new growth. We allow our turf roof to die back completely so that it looks sad and devoid of life in the summer, but watch with glee when the first autumn rains bring it bouncing back to green life again. Obviously, if you choose to grow herbs or vegetables, a more intensive watering regime will need to be carried out. It is necessary to check at least twice a year that no patches in the soil have emerged leaving the membrane exposed and vulnerable to degradation by the sun. If any do develop, cover with more topsoil or cut a piece of turf to patch the hole.

Other roofing options for cob walls

Many old cob buildings with corrugated iron roofs, especially agricultural ones, can be seen all over the country. These have generally replaced the original thatch roof, which would have been removed when they went out of fashion, and because there is a lot less maintenance and expense involved in a metal one. Surprisingly, these roofs can look very attractive and have the advantages of being cost-effective, easy to source, simple to install and maintain, and long-lasting; and the material used can be from a recycled source. Clay tiles also look attractive on cob because they share the same mother material. Clay tiles are common in the south-east of England.

Roofing cob gardens walls

In olden days, when cob garden walls were commonly built, the material of choice for roofing was commonly thatch. They can also be roofed with any of the range of materials discussed above. The wall should have a triangular capping with a roof overhang of at least 300mm (12"), so that water is shed clear of the walls as it falls off the roof. A simple timber frame can be erected to take the roof covering; or a triangle made out of cob, which will cut down on the labour and materials needed to construct a frame made out of timber (see page 101).

Insulation

8

Insulate: to separate something from conducting bodies by means of non-conductors so as to prevent the transfer or loss of heat.
Penguin Concise English Dictionary

Chapter 8

Insulation has been touched on previously in Chapter 2: *Site & design*, in relation to the principles of passive solar design. It is an essential area to get right when building a new cob structure, or indeed any structure that aims to be energy-efficient, in order to maximise the free heat of the sun and to minimise the excessive use of an outside heat source. The concept of insulation relates to the ability of any material to prevent heat or cold from entering or leaving a building.

The issue of insulation is especially pertinent in relation to building a dwelling out of cob. Until recently, building regulations required a target U-value for individual elements of a building, including the external walls of domestic dwellings. This U-value relates to the thermal transmittance of a material, a measure of how much heat a material allows to pass through it over a given timescale. The lower the U-value, the greater the insulation of the material. The old building regulation standard U-value for external walls was 0.35W/m²K, and cob walls on average (cob as a material varies in composition so greatly from place to place that its properties can never be fixed – and herein lies some of the problem!) have a U-value of 0.6–0.7W/m²K. This has been the main point of contention for the passing of new cob structures through the building regulation standards in the past.

Left: Teasing sheep's wool to be used as insulation. Middle: The source of the warmest, most effective and environmentally friendly natural insulation material available. Right: Installing manufactured *Thermafleece* sheep's wool insulation.

The old system of U-value ratings came from Part L of the building regulations (pre-2006). Part L is concerned with the conservation of fuel and power by limiting heat loss through the fabric of the building. In recent years European energy regulations caused the standard U-value rating to decrease, which made it even harder to pass new cob buildings. This came from a growing awareness about the necessity to minimise the use of non-renewable energy sources and the production of CO_2 gases from the burning of fossil fuels.

The new Part L has done away with elemental U-values, and in their place are target U-values, which focus on the thermal efficiency and energy consumption of the building as a whole. This is discussed in detail in Chapter 13: *Planning permission & building regulations*. Although it is still required that any individual element, such as the walls, does not go above a certain U-value (for walls it is 0.7W/m²K), it means that it is now possible to compensate for the high U-values of cob walls by super-insulating other elements in the overall build, such as the roof and foundation plinth, which have low U-values. This makes cob more viable as a modern, 'green' construction material that can satisfy the modern building regulations. Cob more than fulfils the criteria as a sustainable building material. It can be extracted, mixed and built with very little energy, will naturally decompose, and can be recycled over and over again. The challenge is therefore to find ways to use cob that improve its thermal efficiency, so that it can compete with other 'green' building technologies, as discussed in detail in Chapter 13.

Why does cob not have a good U-value rating?

Materials that are renowned for their insulative properties are generally lightweight and have small air gaps in their composition. These air gaps absorb and trap any air that comes into contact with them, thus allowing the air to pass through the material at a very slow rate. This air may be warm or cold, depending on the outside and inside air temperature differences, bearing in mind that warm air is attracted into areas of cold air and vice versa. Materials that have the above properties include, for example, sheep's wool and straw. Cob, on the other hand, is a heavy, compact and dense material. That is why it makes such fantastically solid and durable walls. A well-made cob wall will have very

So what's in a U-value?

W= Watts – the rate of heat loss
M squared = per square metre
K= Kelvin – unit of temperature defined by the Kelvin scale (absolute zero is at 0 degrees, and water freezes at 273.16 degrees)

Right: A deep window reveal showing a typically thick cob wall built for maximum heat retention.

few air spaces in it, apart from the hollow tubes of the straw, which will trap small amounts of air, but not enough to make a considerable difference.

However, despite not having a good U-value rating, cob does have excellent capacitive insulation properties which relates to its thermal mass properties (see Chapter 2, page 37) and its ability to store heat and then release it over a period of time once there is a drop in temperature. For this reason, cob walls have always been built very thick – to maximise the amount of heat they can store, and to increase the amount of time it takes for air to pass through them. This is why cob buildings have always been renowned for being cosy and warm in the winter and cool in the summer. These effects are best felt when a cob building is continuously occupied and being heated from within. Sabine Baring-Gould made a testimony to this in his *Book of the West*, written in the eighteenth century,

with his remark: "No house can be considered more warm and cosy than that built of cob, especially when thatched. It is warm in winter and cool in summer, and I have known labourers bitterly bewail their fate in being transferred from an old fifteenth or sixteenth century cottage into a newly built stone edifice of the most approved style, as they said it was like going out of warm life into a cold grave."

A historical note . . .

As an aside, it is interesting to look deeper into why cob has historically been such a grey area for building control personnel, when there are at least 100,000 ancient, sound cob homes being lived in comfortably today in the UK, some up to 500 years old. To understand this we need to consider how, and by whom, most cob buildings were constructed in the past. As was discussed previously, cob was predominantly the material of the poor, rural tenant farmers, who built by hand from the local soils beneath their feet. The materials at hand, though similar in nature, will have varied in composition enormously from area to area, so it is immediately obvious that trying to standardise such varying materials would be almost impossible. Besides this, there was the stigma attached to cob building – that it was the material of the poor, arbitrarily erected out of *ad hoc* materials. Consequently, until recently, cob was a forgotten building technique, which failed to find its way into the standardised documents of the building regulations. The reluctance of the building control officers to approve its use has arisen more out of a lack of knowledge or definitive information about the material, than from its lack of suitability to produce structurally sound, warm, healthy buildings.

Thankfully, things are changing, and there is a new openness and awareness within the building regulation authorities of cob as a solution to building sustainably. Devon County Council has been a leader in this field, because of Devon's great pride in its heritage of cob buildings, and because of the work done by the Devon Earth Building Association and Plymouth University in carrying out research and tests with cob. This work has helped to earn cob its place as a valuable 21st-century building material, and has helped to ensure that new cob buildings have been and will continue to be approved in greater numbers in the future. Although each local authority differs in its knowledge, approach and attitudes to cob, the precedents set within certain local authorities in Devon

can be utilised for the rest of the UK. Our advice is to establish a good working relationship with a building inspector in your local area, or with an independent approved inspector, with whom you can work to facilitate the successful compliance of a new cob building with the current U-values and all other areas of the building regulations. For more information on cob and the building regulations, please refer to Chapter 13.

The rest of this chapter outlines a range of natural insulation materials that are suitable for insulating various parts of a new cob house. We also talk about ways in which a cob building can be created to maximise the thermal mass properties of cob, and ways to counterbalance its low thermal efficiency rating (U-value), such as a highly insulated straw-bale wall in the north elevation of the building.

Insulation materials: good, bad and ugly

Most conventional insulation materials should be avoided wherever possible in the construction of a building that is made out of natural materials, which seeks to lessen its impact on the environment and be healthy for its inhabitants. There are three main categories of insulation materials, the worst offenders being those derived from non-renewable fossil organic sources. These include, among others, expanded polystyrene, polyurethane foam, polyisocyanurate foam,

Cob and thatch are ideal companion materials, with proven insulation properties difficult to beat even today. Cool in summer and warm in the winter, cob and thatch have a good balance between thermal mass and insulation properties. Builders' details are as on page 25.

and urea and phenol formaldehyde foam. As you might expect, some of these unpronounceable words are bad news for the environment and human health. The next, slightly less offensive category includes materials that are made from non-renewable inorganic sources, but are from a more plentiful source than those listed above. These are made from naturally occurring minerals and include mineral fibres such as rock wool and fibreglass, perlite and vermiculite, and foamed glass. The most eco-friendly and healthy materials are those made from organic materials from renewable sources. These include sheep's wool, flax, cellulose and cork.

The worst-offending insulation materials are so because of their environmental impact and their effect on public health.

Environmental impact
- They require huge amounts of energy to produce.
- The methods used to produce them create CO_2 gases, which deplete the ozone layer.
- They often contain toxic additives, added to increase their resistance to pests, which are harmful to the environment and other living beings.
- Most are not able to be re-used or recycled and will not decompose in landfills.
- They must be disposed of very carefully.

Public health
- Most conventional insulation materials release harmful airborne particles which are carcinogenic, and can cause respiratory problems and skin irritation.
- They are therefore harmful in their

Top: Installing flexible
Thermafleece quilts
in a loft space is an
effortless job.

Above: Loose
sheep's-wool insulation
fitted between timbers.

Centre and above right:
Thermafleece samples showing
quilted sheep's-wool insulation.

Left: Warmcell (recycled
newspaper) between timber
partition walls.

production, application and disposal. Mineral fibre insulation materials are the worst offenders here.

Thankfully, there are some approved alternatives to the above, which are just as, if not more, efficient than conventional insulation materials. These do not harm the environment and are safe – if not a joy – to use and live with. As with all natural building techniques, the art of finding efficient alternatives can arise simply from looking to nature, examining how the creatures keep themselves warm, and also to the methods used by the folk builders throughout the world to protect them from the cold. From here we can adapt and enhance some of the oldest tried and tested methods. There are many ecological options nowadays, but we have chosen to focus on the two most readily available and cost-effective.

Sheep's wool

The first option is inspired from the rugged sheep that comfortably roam the bleak and windswept hillsides of the United Kingdom. Their warm, woollen coats effectively keep in the heat from their bodies, and keep the cold out, by trapping the air in the small air spaces of the wool. Because sheep are abundant throughout the UK, sheep's-wool insulation can always be sourced. It performs as well as, or even better than, equivalent conventional insulation materials, with a thermal performance of 0.037W/mK. This can be compared with foamed glass, which has a measurement of 0.042W/mK, and expanded polystyrene, which has a measurement of 0.34W/mK. It can be used in a variety of

forms – loose or made into quilt form. One such product, which can be purchased in quilt form from a UK-based company called Second Nature, is Thermafleece.

All their wool is produced and sourced in the UK. It is a fully approved product (it has the BBA – British Board of Agrèment – approval), and has the added advantages of being renewable and recyclable. It is produced with a fraction of the energy it takes to produce glass-fibre insulation (14%), and through its production, helps support the economies of some of the poorest rural areas in the UK, as wool currently has a very low economic value.

The advantages of using sheep's wool insulation are as follows:

- It is breathable and can absorb moisture with no loss of thermal efficiency. It is therefore an excellent insulation material for damp British winters.

- It is excellent at condensation control, and actually generates heat when it absorbs this moisture from the atmosphere. This is why we are kept toasty warm by wearing woolly jumpers in the winter. The warmth generated by the wool is not enough to be felt inside the building, but remains inside the cavity space causing the temperature to remain above dew point and hence preventing condensation from forming.

- Wool is naturally fire-resistant. It does not burn when it comes into contact with flames, but actually melts away from the fire source and extinguishes itself. *Thermafleece* is treated with a safe, naturally derived fireproofing agent, which gives it a zero rating for ignitability.

- *Thermafleece* is proofed against insects and pests by being treated with a naturally derived biological control agent. Raw sheep's wool can be treated with borax powder to achieve the same end.

- Sheep's wool is extremely durable and will last in excess of 50 years with no decline in performance.

be fitted in between the rafters on a sloping ceiling, in a loft space, and between internal timber partition walls. It can also be used under suspended floors. In these situations, sheep's wool can replace the conventional glass or mineral fibres, polystyrene and polyurethane insulation materials.

Unfortunately, sheep's wool cannot be used in cavity foundation walls because, being a natural material it needs to have an air gap of at least 50mm (2") to provide sufficient airflow for it to breathe. As far as conventional options go, rock wool is the next best option because it has a low energy

> The most eco-friendly and healthy materials are those made from organic materials from renewable sources. These include sheep's wool, flax, cellulose and cork.

- Quilted forms of sheep's wool insulation come in large rolls, making it extremely flexible and easy to use as it effortlessly takes on the shape of rafters and joists.

- It is completely harmless to humans and animals and can be installed without gloves or protective masks and gear. It will not irritate the eyes, skin or the lungs. In fact, it is a joy to work with, as it still smells sweetly of the sheep, connecting you to the green hillsides from where it came.

In a cob structure, sheep's wool insulation is best used to insulate the roof structure. It can

consumption relative to other insulation materials. Alternatively, you can use vermiculite, perlite or expanded clay balls in the cavity (see Chapter 5: *Foundations*).

Sheep's wool insulation can also be purchased in loose bundles that have been cleaned but not processed into a quilt form. This is a cheaper option, and is best sourced from your nearest available supplier.

If you simply wish to insulate a small garden room or shed, the use of sheep's wool insulation can be taken back to its most raw, natural, unprocessed form.

All that is needed is a trip to a local sheep farmer, where you can purchase large bags of 'unusable' pieces of wool at a fraction of the cost. The lanolin naturally present in the wool should deter moths and pests. You can also use borax powder (see *Resources and suppliers*) or mothballs for added protection. The wool should be teased apart and stuffed into the void between the ceiling and the roof.

Recycled newspaper

When timber is reduced to small particles it has excellent insulating qualities. Cellulose, made from recycled newspapers, has a long tradition in the UK, and was actually patented for the first time in England in 1893. It simply consisted of shredded recycled newspaper, treated with fire-retardant and impregnated with moisture. Today, cellulose fibre is being rediscovered as an environmentally friendly and safe way to insulate buildings.

There is currently a product on the market called 'Warmcell 100', which comes in either an easy-to-lay quilt form or in a form that is blown in by a specialist installer. It is best suited to loft spaces, suspended floors and timber partition wall systems. It can be used in place of glass and mineral fibre insulation. It is however necessary for the roof to have a felt membrane, laid under the slates or tiles as extra protection against water penetration.

Its positive attributes include:

- It is extracted from 100% recycled news-papers, and therefore gives a new form of life to this material.

- It is non-toxic, non-irritant, and is not hazardous to health.

- It can be disposed of safely without creating toxic waste or biodegradability problems.

- It requires very little energy to manufacture.

- When blown in, it forms a complete blanket with no gaps, cracks, or bridges for heat to escape, unlike quilt insulation materials.

- It has excellent thermal conductivity of $0.033W/m^2K$.

- It is completely fire-resistant, and meets the British fire protection standards through the addition of simple and safe inorganic salts.

- It is safely treated to protect against fungus and insects.

In a cob house, cellulose fibre insulation is best used in a loft space or timber partition walls, where it can be sprayed or laid in place.

Other alternative insulation materials

There are many great products on the market including those made from hemp, flax, wood wool, wood fibre, cork, and compressed straw, which are all good for various uses throughout a building. In some instances however, this is one area of the building where it may be justified to use a more conventional product. When calculating the amount of embodied energy involved in the production of a material, there are many elements to take into consideration.

A hybrid cob building – using the inherently excellent insulation values of straw bales in the cold north walls. Straw bales must be stepped and anchored into the cob wall.

Among these are the amount of energy consumed in the transportation of the product, its effectiveness during its life, and its life expectancy. In some cases, a more conventional insulation product may be a viable option even though it has used a lot of energy to produce, because it may be sourced more locally than some of the natural insulation materials, may be more effective (for example in the foundations), and last longer.

We recommend avoiding all those products derived from non-renewable fossil organic sources, but do accept that there is a definite place for those products derived from naturally occurring minerals that are non-renewable but plentiful.

Improving thermal efficiency

To improve the overall thermal efficiency of a cob house, and to compensate for cob's inherently high U-value, there are some simple measures that can be taken. These include methods to increase the insulation on the cold, north wall of the building, and utilising cob's ability to absorb and release heat because of its good thermal mass properties. These can be used in conjunction with the principles of passive solar design (see Chapter 2: *Site & design*). The following solutions should also be considered in conjunction with the more encompassing list included in Chapter 13.

Straw bales on the north wall

Straw-bale building is a natural building technique that is rapidly gaining in popularity in many parts of the world. The process of building a straw-bale wall involves using rectangular straw bales as large building blocks, stacked on top of one another, compressed and pinned. They are then covered with earthen plaster sub-coats and finished with a lime render/plaster.

Straw-bale walls provide excellent thermal insulation, with a U-value of 0.13W/m²K – far lower than is required for the standard building regulation. A straw-bale wall can be incorporated into a cob building to make up the cold, north wall (because due north doesn't receive direct sunlight), and hence increase the overall thermal insulation of the building.

It is essential to ensure that both sides of the straw bale wall are solidly tied into the cob walls using ties, to avoid any structural faults between the two. This can be achieved by stepping the bales at every other course so that you avoid creating a vertical seam, which would leave the two walls vulnerable to separation in the future. This is especially pertinent for cob, because it has a natural tendency to shrink slightly as it dries.

It is also important to ensure that the wall plate junctions do not end up occurring directly above the meeting point of the straw bale and cob walls. As long as this is avoided, the wall plate will act positively to tie the walls together from the top.

A house built totally of straw bales would obviously provide excellent all-over thermal insulation. However, straw bales lack the excellent thermal mass properties of cob, and straw bales are not as flexible as cob in creating highly unique shapes and forms. Bringing the two together in this way will certainly ensure that the two materials are utilised in a way that maximises the positive attributes of both.

For a detailed guide on how to build with straw bales in the UK, consult Barbara Jones's excellent book *Building with Straw Bales*, or pick up Bill and Athena Steen's book *The Straw Bale House*.

Insulated north wall bookshelf

This option works well in a small house, and involves creating a wall to ceiling bookshelf against the north cob wall, with an insulated cavity between the wall and the back of the bookshelf (e.g. using sheep's wool or cellulose fibre insulation). If the bookshelf is filled with books, this will provide an extra skin of insulation as well.

A note on windows

Windows are a potential area of major heat loss from buildings. We therefore encourage you to source the most thermally efficient windows that you can afford – they will pay for themselves with the amount you will save on your heating expenditure. These should be at least double-glazed, low e-value panes. Triple-glazed are best. The windows must also be fitted tightly to prevent heat loss through air gaps.

Lime & other natural finishes

In contrast to modern buildings, traditional cob houses with thatched roofs are in a very real sense 'organic', and like any living organism must, to remain healthy, be able to 'breathe' and to respond readily to variations in temperature and humidity.

Larry Keefe

The role of a plaster or render finish on a cob wall is many-fold. On the outside, render serves as a sacrificial element, protecting the walls from driving rain and wind, and on the inside plaster serves to prevent dusting from the cob, to lighten the interior space, and to provide a background for decorative natural paints and pigments. A render or plaster specifically applied to any sort of earthen wall must serve the purpose of allowing the cob to breathe, and must therefore be made of a porous material. Lime and earth are the preferred materials of choice as they are soft, breathable and flexible, and will truly enhance the forms and curves inherent in the cob. Different cultures around the world use many varying techniques and mediums to protect their earthen walls, but the principle ingredients are normally derived from either earth or lime, and the unifying characteristics are that they are soft and breathable. In this chapter we will give a breakdown of the main breathable finishes that are most compatible with cob buildings, including basic information, recipes, preparation, application and the tools you will need.

The principal breathable finishes that we recommend for cob walls are lime, earth plasters, earth/lime plasters, lime wash, natural breathable paints based on plant materials, alis clay paints and casein paints. A freshly built cob wall will take at least one year to fully dry out and complete its major

shifting and settling. For this reason, we recommend that external rendering should take place after this period to prevent major cracking in the lime and to enable maximum evaporation of moisture. The inside, however, can be plastered from three to six months after completion of building.

Why do cob walls need to breathe?

To remain healthy and last for centuries, cob walls must be able to breathe. All lime and earth renders, plasters and washes are porous, and therefore allow moisture to pass in and out of the walls. The application of waterproof, non-porous materials such as cement will prevent this exchange of moisture and will eventually cause a cob wall to fail, for three reasons:

1 The golden rule for traditional buildings is to use like-for-like materials. Cement is hard and brittle and therefore does not work when put on cob, which is soft and flexible. This means that a cement finish will inevitably crack, because it cannot cope with the small movements of a flexible cob wall. These cracks may allow moisture into the wall, but the impervious cement covering will not allow it to escape, which may eventually cause it to trickle down into the base of the wall and manifest problems such as dampness or eventual collapse.

2 Moisture entering the cob wall due to other defects in the building such as a faulty roof or gutter, will get trapped behind an impervious material such as cement and again, at best will lead to dampness in the wall, and at worst will cause the cob wall to literally melt from within, leading to collapse if left unchecked for a long time. Cement renders can also mask underlying problems in a cob wall.

3 Cob walls do need a certain amount of moisture within them to remain healthy, so that they do not become bone dry. We have removed cement renders from old cob walls that have successfully kept the moisture out, but left the wall dry and very crumbly underneath, and therefore unable to hold its form. To help you visualise the implications of covering a cob wall with cement, imagine yourself being covered from head-to-toe in plastic. Like us, all natural building materials including wood, stone and grasses, need to breathe to remain healthy.

As well as the structural problems mentioned above, a build-up of moisture that cannot escape will also cause moulds and fungi to develop, which are not only damaging to the health of the building but are also detrimental to the health of the occupants. This also applies to other non-breathable finishes, including tar and plastic paints.

Does your cob wall need a protective coating?

Cob walls can be left un-rendered as long as they do not receive driving rain, or experience prolonged freezing conditions. Over time, these will lead to surface erosion, though it will take a long time to reach a point where the wall is jeopardised. If you enjoy the look of the raw cob finish, spend a winter observing how heavy rainstorms or the freeze-thaw affect it, and apply a protective finish later if necessary. Right: Exposed earthen test wall showing the effects of severe wind and rain.

Lime finishes

In the UK, lime has been used for hundreds of years as the natural partner to cob walls. The use of lime to make plasters and mortars dates back to about six thousand years ago. Up until the mid-nineteenth century, when cement and gypsum came onto the scene, the practice of producing, preparing and applying lime finishes was a part of local, vernacular practice.

The annual spring clean always involved the application of a fresh coat of lime wash, not only to spruce up the house but also to prevent the spread of germs through lime's antiseptic qualities. In this section, we hope to describe how simple it is to prepare and apply lime finishes, and to share with you the many benefits – environmental, social and practical – that using these age-old techniques will bring to contemporary buildings.

Why use lime?

Environmental credentials

Lime is better for the environment than cement for three main reasons:

1 Cement has a much higher 'embodied energy' than lime. This takes into account the energy used at all stages of a materials production, and disposal, including: the energy used to extract the raw materials, the transport to the processing plant, the energy to process the raw materials, the transportation to the site where the material is going to be used, the energy needed to install the product, and the potential energy needed to dispose of the product when it is no longer needed in the building. Cement is also cut with many toxic chemical additives that are potentially destructive to the environment.

2 Cement contributes significantly to the greenhouse gas effect, because during its production it releases large amounts of carbon dioxide into the atmosphere. Lime, on the other hand, despite also releasing carbon dioxide during its production, is able to reabsorb the carbon dioxide that has been released during its manufacturing, when it sets on the wall.

3 Cement is a one-way material, meaning that once set, it is not able to revert back to its original raw form; hence it will not biodegrade, and cannot be used again. Lime, on the other hand, will biodegrade back into the earth, and convert back into its raw form – calcium carbonate. Old lime putty plasters and renders can even be removed, soaked, re-mixed and put back on the wall.

Lime is compatible with earth

Lime bonds extremely well to earth because it is an equally soft and porous material. Cement, which is hard and brittle, bonds very weakly. This means that a cement render may separate from a cob wall causing it to fall off in large chunks. Lime is a good partner to earthen walls, because lime and earth contract and expand at the same rate, which means that the lime finish will remain stable and well adhered.

Lime is good for our health

Due to lime's high alkalinity, it is antiseptic, anti-bacterial and anti-fungal, and is therefore a very hygienic wall finish, especially on internal walls. It also has the ability to absorb and hold moisture during periods of high humidity, and then release this moisture when relative humidity levels drop. Lime is therefore called a hygroscopic material, and will balance moisture levels in the atmosphere. In an inside space, this will help to prevent condensation, the main cause of moulds and fungi. For these reasons, lime wash was traditionally used to paint the inside of cattle pens, to help prevent the spread of disease amongst the animals. Most modern chemically-based materials, such as oil-based and acrylic paints, can produce nasty vapours through off-gassing even after they have been applied and dried. The exposure to these toxins in our indoor environments has been linked to a vast spectrum of cancers and respiratory illnesses; so much so, that the concepts of 'sick building syndrome' and 'environmental illnesses' are now very much a part of our vocabulary.

Lime has great aesthetic values

A wall finished with lime looks soft and beautiful. It gives a matte, chalky finish, and will accentuate and highlight the sensuous curves of a cob wall. A wall finished with lime will age well and blend into the natural environment. A cement finish, in comparison, will generally give a hard, rigid look. A finish lime wash with a natural pigment can be used to create unique, earthy colours, which will be dynamic, constantly changing with the light, time of day and the weather.

Lime sets more slowly than cement or gypsum

Lime is more forgiving than cement in the application process. It takes a lot longer to set than cement, and can generally be re-worked

Left: Before the Industrial Revolution, the production of lime was a local affair, and most communities would have had their own lime kiln like this disused and overgrown one in a village in Surrey.

Right: Lime putty maturing at the Cornish Lime Company, Bodmin.

if it is wetted down for up to 24 hours after application. This allows the novice plasterer the chance to re-work any mistakes, and to work slowly before the techniques are mastered.

The application of lime requires sensitivity to the weather and seasons

Cement can be applied in all weathers and seasons, whereas lime, to get the best results, can only be applied within a certain set of conditions (see next page). This encourages the fostering of a deeper relationship between you, the building, and the environment in which it is situated.

Lime has the ability to self-heal

Lime has the ability to self-heal any small, hairline cracks that may appear immediately after applying the plaster or render. This is because when the carbonation process occurs as it sets on the wall, (i.e. the calcium hydroxide in the lime reacts with the carbon

dioxide in the atmosphere). Calcium carbonate is produced, which fills the small cracks. We also find that small cracks can be filled successfully with the slurry from the bottom of the lime wash bucket.

The production of lime and the lime cycle

The raw material for making lime is calcium carbonate, which occurs naturally in limestone, shell or coral. This is heated in a kiln to very high temperatures reaching $1,200^{0}$C. As it heats up, steam is driven off and then carbon dioxide, and a chemical change takes place, leaving calcium oxide in the form of white lumps. This is known as lump lime or quicklime, and is the raw material for making lime putty, which is what you will be using with sand to render/plaster your cob wall.

The process of making lime putty from the calcium oxide is called slaking, which

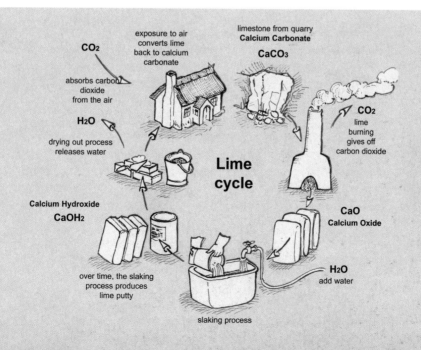

Lime cycle

limestone from quarry
Calcium Carbonate
CaCO₃

CO_2
lime burning gives off carbon dioxide

exposure to air converts lime back to calcium carbonate

CO_2
absorbs carbon dioxide from the air

H_2O
drying out process releases water

CaO
Calcium Oxide

H_2O
add water

Calcium Hydroxide
CaOH₂

over time, the slaking process produces lime putty

slaking process

involves adding water to the quicklime. You can purchase quicklime and do this yourself at home if you really want to go the whole hog, but it is a dangerous process which can burn your skin on contact or blind you if it gets in your eyes. It is much easier and safer to buy ready-prepared and matured (it must be matured for a minimum of four months – the longer lime putty sits around, the better quality it becomes) lime putty from a quality supplier here in the UK.

Different types of lime available

There are three different types of lime that you can purchase to make lime finishes. These are:

- Lime putty
- Natural hydraulic limes (NHL 2, 3.5 and 5)
- Hydrated or bagged lime

Lime putty

Lime putty was traditionally the basis of all lime plasters, renders and washes, and is still by far the most suitable material to use on cob walls. This is because it is a purer material and much softer. It is therefore more porous and breathable than the natural hydraulic limes or hydrated lime, and will bond very well to cob. It comes in large white tubs and resembles thick yoghurt.

Natural hydraulic lime

The next best option is to use natural hydraulic lime. This comes in a white powdered form and is created from limestone that contains impurities such as silica and alumina. These impurities are present in varying amounts, ranging from 8% to 25%. The processed lime is graded according to how much impurity is present. NHL 2 contains less than 12% impurities, NHL 3.5 up to 15%, and NHL 5 up to 25% impurities. The amount of impurities present determines the strength and permeability of the lime.

NHL 5 is the strongest and most impermeable to water, and will achieve a set and finish closest to cement. NHL 2 is the weakest, and is closest in strength and permeability to lime putty. The main difference between pure lime putty and natural hydraulic lime is that the latter will set under water, and hence is known as 'hydraulic' lime. Lime putty needs to be exposed to air to achieve a set, and sets much more slowly. NHL limes can set under a range of wet conditions, which is why NHL 5 is used most often for building bridges, dams, harbours, for rendering in damp climates, and for extremely exposed weather walls. It is also a good alternative to using cement for stonework, as well as for making 'lime-crete' foundations (see Chapter 5).

Hydraulic limes were used for these purposes before the invention of Portland cement, and were home-produced in the UK, mainly from the Blue Lias limestone of southern England. Today, the best quality natural hydraulic limes are sourced from Europe, mainly France, and a product that we have successfully used is St Astier natural hydraulic lime. Even though we recommend using lime putty rather than natural hydraulic limes, we have also used these successfully with cob buildings.

NHL 2 works well as a lime wash, NHL 3.5 for internal plastering and external rendering of cob walls, and NHL 5 for the purposes mentioned on page 140. They are suitable in less favourable climatic conditions, and on walls that are excessively wet. However, due to the impurities present, the permeability of the natural hydraulic limes (NHL) is lower, and they are therefore less breathable. Hydraulic limes have a higher environmental impact than lime putty, as they only absorb around 50% of the carbon dioxide emitted during their manufacturing. However, they still have 30% less embodied energy than cement. One must also consider that good hydraulic limes are imported from Europe, which means transport implications and costs, whereas lime putty is made throughout the UK (see *Resources*) and is, in our opinion, a more suitable, appropriate and nicer material to work with.

Bagged or hydrated lime

Bagged or hydrated lime powder (made by Blue-Circle) is the lime that you find in general builders'-merchants. We do not recommend that you use it for plastering or rendering on top of cob. It is far inferior to the natural hydraulic limes and lime putty, and it has generally been sitting around for some time because there is not so much demand for it in the conventional building world. Powdered lime that has been sitting around for some time will have deteriorated, and possibly partly gone off due to the exposure to moisture and air. If this is your

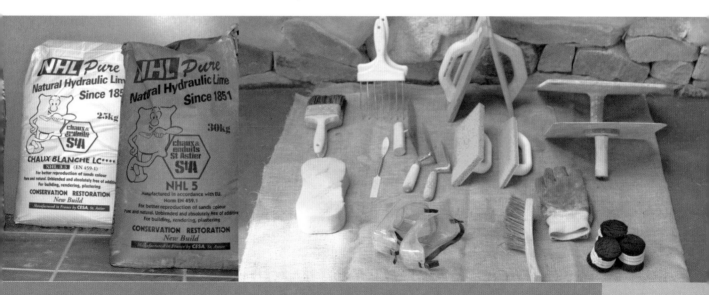

Above: Natural hydraulic lime 2, 3.5 and 5 can be bought from your local lime supplier in bagged, powdered form.

for making & applying lime finishes

1   Selection of trowels and floats
2   Hawk
3   Water mister or old paintbrushes
4   Scratching tool
5   Buckets

tools

Standard lime plaster and render mix

The standard lime plaster/render mix for all three coats uses: one part of lime to three parts of sand to quarter part of fibre (fibres are optional for the final coat).

Mixing Lime
Making a lime mix with lime putty

Lime can be mixed by hand in a wheelbarrow using a shovel or a larry (a tool that looks like a garden hoe with holes in it) when doing small jobs. For larger projects, a standard drum cement mixer, or barrel mixer, is a better option.

1 Add lime and sand to mixer
Turn on the mixer and add alternate shovels full of lime putty and sand. One shovelful of lime to three shovels of sand – it may seem an obvious point, but make sure your shovel-loads are roughly the same size. Keep on adding the lime and sand until the mixer is full but not brimming over. If you let it brim over, you will lose half your mix and the machine may topple over.

2 Add fibres
Fibres need to be individually teased into the mix and not all thrown in at once. The latter method will cause the fibres to clump into a ball and prevent them from being evenly distributed throughout the mix. Unless the sand is bone dry, do not add any water – the moisture in the lime putty should be sufficient. Rotate for at

least fifteen minutes and if the mix seems dry and the sand and putty are not mixing together thoroughly, or they start to ball up and get crumbly, gradually add small amounts of water.

3 Allow mix to rotate
Allow the mix to rotate for at least twenty to thirty minutes. When making a lime putty/sand mix in a mixer it takes a lot longer to achieve the desired results than when making a sand/cement mix. Allowing the mix to rotate for at least twenty minutes will help achieve the wonderful creamy consistency that makes for an easily workable render/plaster.

The final mix should be the consistency of stiff cake batter. It should be able to hold its form without slumping. If it is at all sloppy, you have added too much water and you should let it sit for a while to dry out, or compensate by adding more dry sand. If the mix is too wet on application there is a higher risk of the render/plaster shrinking as it dries out, which will lead to cracking. It will also be very hard to work with.

In an ideal situation, this mix should be stored for a few weeks in sealed buckets or wrapped up in a tarp and knocked up in the mixer for about ten minutes before applying. It is not essential to do this but a stored mix will improve over time, making it easier to work with and potentially giving better results.

A trowel full of lime scratch coat (left) and finish coat (right) showing ideal consistency

only option, soak the powder in water for up to one week to create putty.

Preparing lime putty renders and plasters

A render refers to the protective lime coating applied to the outside walls of a building, and a plaster refers to inside applications. Generally the same mix is used for the outside and the inside. Lime is a joy to work with, but take care when handling: always wear gloves, as it is a caustic material and will burn your skin on contact. If it gets on your hands, wash immediately with soap and water and rub your hands with vinegar – the acid in the vinegar will neutralise the alkalinity of the lime. Similarly, if it gets in your eyes wash out immediately with clean water or a saline solution.

The ingredients for a general lime plaster or render mix

The three basic ingredients that you will need in different proportions are:

- Lime putty or NHL 3.5 lime
- Sand
- Fibres

Lime

You should try to obtain the best quality lime you can from a reputable supplier, to ensure you get the best results. The next most important ingredient is the sand you use.

Sand

Make sure that the sand is well graded, and has sharp, varying grain sizes. You can generally purchase sand from your lime suppliers, or ask their advice on where good, local sands are produced, and which works best in your region. The role of the sand in the mix is to reduce shrinkage and cracking. The principles are similar to the cob mix, with the lime acting as a binder for the sand particles. Generally, even though you can achieve a smooth finish with coarse sand, we always use fine sand in our topcoat, which produces a smooth creamy finish.

Fibre

The role of the fibre in a lime plaster or render is again the same as the role of the straw in a cob mix – it gives the mix tensile strength, enabling the plaster/render to cope with the microscopic movements in a cob building. It also helps to control shrinkage and prevent cracking. The fibre should be clean, and roughly 6–19mm ($1/4"$ –$3/4"$) long. Traditionally cow hair was used, but this is hard to come by these days, and unless you live near or around animals it is easiest to obtain fibre from your lime supplier. The hair of deer, yak, llama, ox, goat or horse can be used. We also use polypropylene fibres, which are a by-product of the rope industry. We always use fibres in our scratch coats but often will refrain from using them in our final coat to allow for a smooth finish surface.

Optional addition of sieved earth

If you would like to maintain the earthy colour of the cob in your final finish, you can add finely sieved clay to your lime/sand mix. This addition means that the mix is liable to crack more easily, but this can be avoided by adding more sand.

The Layering System

When plastering or rendering a cob wall with lime, there are generally two to three different coats applied, which are built up in layers:

1 Dub coat: filling in any large recesses in the wall. This only really applies to the restoration of old cob walls. Hopefully your new walls will be in a good enough condition to forego the dub coat.

2 Scratch coat: this is the first full coat to be applied to a cob wall. It is called a scratch coat because it is 'scratched up', to provide a key for the next coat. This coat helps to iron out any protrusions or small recesses in the wall to create a smooth surface for the topcoat.

3 Final coat: this is the smooth finish coat onto which the lime wash is applied.

4 Lime wash: this is the final, protective and decorative lime paint finish, which can be coloured with natural pigments.

Making a lime mix with natural hydraulic lime (NHL 2 and 3.5)

We have had good results using both NHL 2 and NHL 3.5 on cob walls. Use the 3.5 where there is more exposure to the weather, such as driving rain.

Mixing with NHL limes is exactly the same as making a mix with lime putty, only more water is needed because the NHL comes in a dry powder form. Another key difference is that because NHL limes set under water and go off much quicker, you must omit the storing stage described above. The NHL mix must be used immediately before it sets hard in the bucket or wheelbarrow, but it still takes longer to set than cement does. Wear a mask when shovelling NHL lime powder into the mixer, as the fine particles will fill the air and are not good to inhale.

Using a pre-mixed lime/sand mix

There is another option you can take, if time is limited and you lack the equipment needed to make a lime/sand mix. You can order, and have delivered, perfectly mixed up lime/sand mixes ready to be applied onto the wall. You can order the mixes either in ten gallon buckets or by the tonne in large sacks (see Suppliers list).

Preparing surfaces to be lime rendered or plastered

When preparing the area to be rendered or plastered there are three things to consider:

1 Lime rendering and plastering can be a messy job, especially when you are new to it. Be sure to cover any areas that you want to avoid getting covered in lime, as lime is difficult to clean from stones and wood. Wash off immediately with water and a hard brush if you have any accidents. Any lime that dries onto stones can be brushed off with a wire brush. To avoid wastage onto the ground, place a clean tarp or piece of plywood

at the base of the wall to collect any droppings which can be remixed and used again.

2 The cob wall surface that is to be rendered or plastered must be dampened before the lime is applied, so that there is a good suction bond between the lime and the earth. For this purpose, a simple paintbrush dipped into a bucket of water and flicked or brushed onto the wall works well. Equally effective, and especially good on friable cob walls, is a fine mist sprayer, which can be purchased at a garden centre. This can also be used to keep the finished plaster or render moist after application. Keep mucky lime hands away from the nozzle and trigger if you want it to last for more than one day!

3 Assess your scaffolding needs before you begin working, and make sure you have the right ladders and scaffolding available, to enable you to work safely and at ease on the higher reaches of your walls.

Application

1 The first coat on new cob walls is the scratch coat. As with all these techniques, there are many different ways to apply lime, and over time you will probably come up with a way that suits you best. When starting out, the easiest method is to throw the lime by hand onto the wall, and then trowel it in. Alternatively, you can use a hawk to hold the render/plaster, which can be scooped off the hawk and applied onto the wall with a trowel. If you find that your arm gets tired holding the hawk, you can simply scoop up globs into your free, gloved hand and apply

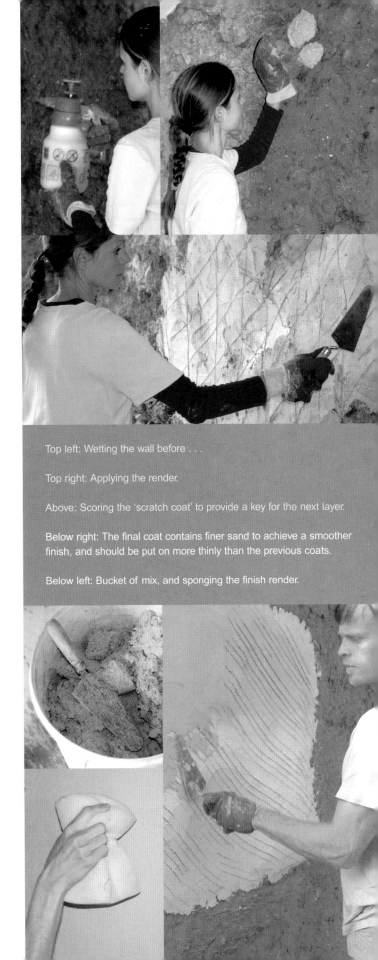

Top left: Wetting the wall before . . .

Top right: Applying the render.

Above: Scoring the 'scratch coat' to provide a key for the next layer.

Below right: The final coat contains finer sand to achieve a smoother finish, and should be put on more thinly than the previous coats.

Below left: Bucket of mix, and sponging the finish render.

Applying lime plaster/render

1 Dampen the cob wall and forcefully throw globs of lime onto it.

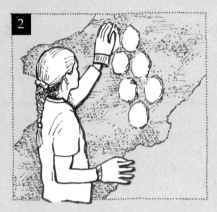

2 Throw the lime until a group of about six globs are on the wall.

3 Using a trowel, smear globs together, working them forcefully into the wall.

4 Once the wall is covered, create scratch marks in the lime using scratcher tool or trowel as a key for the next coat.

5 Apply a thin layer 3–5mm ($1/8$"–$3/16$") thick on top of dampened, set scratch coat.

6 When almost set ('green hard'), sponge or hard trowel to create a smooth finish.

A hawk and trowel can be used instead of the throwing method, but throwing provides the best adhesion onto the cob especially for older walls with a crumbly surface.

Use a lot of force in smearing the lime into the wall so that really good contact is made between the cob and the lime.

When scoring the surface make a shallow scratch and not a deep cut through the full thickness of the lime.

The lime is best applied with a hawk and trowel for the final coat. Again, it needs to be pressed forcefully into the wall, so that it fills the voids created in the scratch coat, and it is keyed into the surface of the wall.

The final coat should be hard trowelled or damp sponged when it is hard enough to hold its form, but soft enough to be worked. Trowelling creates a slightly shiny finish, whereas sponging creates a softer finish.

the lime direct to your trowel from this hand. Always start at the top of the wall and work downwards, so that you don't drop lime onto your finished work. You can work from left to right, or right to left – whatever suits you best. Find a rhythm that feels comfortable, and stick with it. Start with a small metal gauging trowel, which will give you a good feeling of control. Once you have perfected this method, you can try using a large, steel or wooden plastering float. If you are working on an uneven wall surface, shortening the float at either end is essential for making the application easier.

2 Make sure the wall is sufficiently damp, but not dripping wet.

3 Grab a handful of render/plaster with your hands, and pull away small globs, which are then thrown forcefully (but not so hard as to splatter everywhere; be sure to take care of your eyes – wear goggles!) onto the wall in groups of three or four to cover a small area. The forceful action of throwing the lime onto the cob will produce a superior bond than is achieved by trowelling it on.

4 Blend and smear lime into the wall. Use a trowel to blend the globs together, smearing them into one unit over the wall surface. Press hard into the wall with the trowel, smearing backwards and forwards, so that good contact is made between the lime and the cob. Achieve a thickness that is no more than 16mm (just over $1/2$"). Any more than this, and your render/plaster is liable to crack. In your scratch and dub coats you can

afford to go a little thicker, as a small amount of surface cracking will not harm your finish coat. The undercoats provide the perfect time to experiment if it is your first time using lime.

5 Continue with this method until the whole wall is covered.

6 When you have covered the total area of wall that you want to for the day, make score marks with a scratching tool. This is to provide a key for the next coat, and will also help to regulate cracking, as the cracks will be more liable to follow the score lines. You can achieve this with a comb scratcher, which can be bought from a lime supplier or your local builders'-merchant. Do not score too deeply into the lime surface. Allow the render to go off for at least three to four days.

7 Dampen the wall and apply the final coat using a trowel and hawk, and allow the plaster/render to go off so that it is 'green hard' (will not dent when you press your finger against it). The final coat must be thinner than the scratch coats – from 3–5mm ($1/8$"–$3/16$") thick – as the finer sand used in the mix will make it more prone to cracking.

8 At this point, you have two options to create your finished look: for a smooth, slightly shiny finish, you can use your trowel (or a wooden float if you are working on a fairly straight wall) to make small, circular movements, and rework the wall after finely misting it down. This should be done when the render/plaster is green hard, which will vary according to the weather/environment.

Alternatively, for a slightly grainier finish, again wait until the render/plaster is 'green hard', and use a dampened sponge worked in circular movements over the surface. This method is great for feathering in join lines, and for losing any small cracks that may have cropped up. It is also a more effective method when you are working on an undulating cob wall with lots of lumps and bumps.

9 Keep wall damp while lime sets. Fine mist the wall for regularly, for at least 48 hours after completion of the final coat, as the longer the lime takes to dry, the harder the set. This also enables you to re-work any small cracks that may have appeared, either with a sponge or a trowel, depending on which method you used to finish the coat. You are now ready for the really rewarding and fun part of the process: lime washing and adding colour to your beautifully hand-crafted wall.

Some general points to note whilst applying lime:

- Never render in the rain. If there is a threat of rain when you have completed your work, make sure your wall is protected with tarps for at least three days after completion.

- Never render during cold or freezing weather, or if this type of weather (5°C or lower) is imminent.

- Never apply the render/plaster thicker than 16mm (just over $1/2$") when applying lime, as it will simply crack and may need to be re-done. A thick layer of lime prevents the required contact with carbon dioxide in the atmosphere (which is reabsorbed by the lime during the carbonation process to create the set). Also, a thick coat will have a reduced bonding with the wall, and the weight of the lime may cause it to pull off the wall.

Applying lime with a render gun

The use of a render gun significantly speeds up the process. Typically, you may be two to three times faster than plastering by hand. Additionally, the force at which the render comes out of the gun creates a great bond between the lime and the surface of the wall. This is a particularly good method to use on old crumbly cob walls. The negative side to using a gun is that you are burning diesel, a precious fossil fuel, putting pollution into the air and creating some obnoxious noise – it's one of those compromises again.

What you will need

A Sabion render gun works very well. This can be bought from the Anglia Lime Co. This gun originates from France, where the application of lime is often carried out with a compressor and gun.

You will need an air compressor with 35cfm at 50 psi (pounds per square inch), which will be connected to the gun by hosing pipe. You can rent the air compressor from your local tool hire centre. You should request a trailer-mounted 'single tool' compressor, set at the above specification.

Application

Cover all windows, doors, floor and timber with plastic or tape before starting work, to protect them from lime splatter.

The wall should be dampened and prepared in the same manner as described earlier.

1 Mixes should be ready-made and waiting in buckets. There is an art in getting the render just the right consistency for the gun – it has to be wet enough to spray on easily without clogging the gun, but not so wet that it does not hold its form on the wall. Another handy tip is to use very short fibres in your lime mix, as long ones can block the exit points of the sprayer. This can be very frustrating and will hold up the flow of progress considerably.

2 Fill the render gun hopper (headpiece) with the lime mix. Stand roughly a foot away from the wall, pull the trigger, and fire the lime onto the wall. It is very important to wear protective gear, i.e. goggles, gloves, hat or hood to protect hair and face, a jacket to protect arms, and cover all exposed skin.

The power of the gun is quite strong, so start out by getting to know it and doing some test patches. Notice the amount of power relative to the pull on the trigger and how different this feels, e.g. halfway pulled compared to fully pulled. When you are doing more detailed work such as around a window, it is much easier to control when the trigger is only half pulled back.

3 Trowel in or leave rough for desired finish. Once sprayed onto the wall, the lime can either be trowelled in to create a smooth finish, or simply left to create a rougher, harled appearance.

Selecting the different attachments

For a scratch coat: use the three-holed attachment so that you get a wide and large spray. Apply the render liberally, covering all areas. Do one section at a time 2.5 x 2.5 metres (8 x 8 feet), then stop spraying to hard trowel

Above left and right: Using the Sabion lime render gun greatly increases the speed at which lime is applied to the wall. However, he should be wearing goggles!

Middle: The render gun apparatus, with air hose and spray nozzle fittings.

the sprayed lime, if this is the finish you choose. While you hard trowel it, scrape off any excess render that has been sprayed on, and re-apply it in areas that look a bit thin. For a final coat: use a smaller-holed attachment so that your spray is more controlled and contained. The same rules detailed above apply, but do not spray so liberally. Focus on covering the surface with a uniform amount of lime. Again, scrape off any excess plaster/render, and redistribute in areas that are too thin, or simply put the excess lime in a bucket for the next stretch of wall.

Maintenance of the gun

The gun needs to be thoroughly cleaned after each use, as otherwise the lime will dry and render the gun unusable! A toothbrush works well for the nooks and crannies. Always fully submerge the hopper into a bucket of clean water and pull the trigger, whilst connected to the compressor, to clean out the nozzle heads. Every so often, wipe the gun over with some light oil or diesel, to keep it in good condition, and keep the trigger well lubricated.

When applying lime to a newly-built cob structure, it is best to wait for at least one year before rendering the outside, because there will still be some minor movement going on in the building as it dries out. The inside, however, can be plastered after a few months of drying time.

Tyrolean machine

A tyrolean machine is a hand-powered contraption which produces a finish that is half way between a render/plaster and a lime wash, known as a 'spatterdash' finish. A slurry of sand, hydraulic lime and water in the proportions of 1:1:1 or 1:2:1 is added to the machine, which is then propelled onto the wall surface by means of cranking a handle to turn a flicker inside the machine. The finish produced is similar to the tradi-tional rough cast – slightly jagged and spiky. It is especially good on very old, friable cob walls, which are fragile to the touch.

Earthen finishes

Around the world, earthen finishes have been used amongst all cultures that have a tradition of earth building. In the adobe tradition of New Mexico, USA, there is still an annual spring community gathering to maintain and repair the earthen plaster on the historic adobe church of St Francis. Similarly, in Mali, Africa, every spring a competition is held between two halves of the town to re-plaster the great mud brick mosque of Djenne.

In the UK, earthen plasters – plasters or renders made out of sieved clay subsoil, sand and fibre (chopped straw or animal hair) – were traditionally used internally and exter-nally on cob or stone buildings up until the 19th century. In most cases, they were used as a base coat for a lime plaster or lime wash. We feel that earthen plasters have a very definite place in natural building today, especially on cob buildings. There is nothing more satisfying or rewarding than using your

A crew of community members re-rendering a living compound in Timbuktu with earth render.

bare hands to smear on this sticky, smooth paste to create a beautiful soft and sensuous finish to your cob wall. As opposed to lime, an earthen plaster is completely safe and non-corrosive to your skin, and is therefore excellent for children, and for beginners to plastering. Earthen plasters are also extremely forgiving as they set very slowly, allowing a relaxed work pace, and can always be wet down and reworked if mistakes are made. In fact, for undercoats, the rougher the finish means the better the key for the attachment of the final coat of plaster. As with cob, the environmental impact of extracting, making and applying earthen plasters is incredibly

low – again, the clay can be extracted from your own backyard with your hands and a shovel and can be mixed by foot. They can be inexpensive, your only costs being the purchasing of sand and straw, and your time spent extracting the clay, mixing it and then applying it to the wall.

In the British climate, especially on weather-facing walls with driving rain, we do not recommend using earthen plasters externally without several coats of lime wash. Earthen plasters are ideally suited to internal or well-protected spaces, or as a base coat underneath a finish coat of lime plaster or render.

Test batches and recipes

Making an earthen plaster is very similar to making a batch of cob. Your recipe will depend very much on the properties of your local clay subsoil, and will be best achieved through carrying out at least four test batches applied onto a section of protected cob wall. (For mixing and applying your test batches, refer to next section in this chapter.)

Start with: 1 part clay subsoil : 1 part sand : $1/2$ part dry chopped straw in a bucket.
For other test batches, try adding more sand:
1 part clay : 2 parts sand : $1/2$ part fibre.
Try adding more clay:
2 parts clay : 1 part sand : $1/2$ part straw.
Try a high straw mix and see how it holds up:
1 part clay : 1 part sand : 1 $1/2$ parts straw.

Sieve your clay through a 13mm ($1/2$") screen to get rid of any small stones that may get in your way when plastering. Play around with different mixes until you get one that works well. As you apply the mix, ask yourself whether the plaster goes on smoothly. If not, you may need to make it creamier by adding a little more water to your mix. Allow the plaster to dry for at least a couple of days on your test wall before carrying out your assessment. Your assessment will involve asking the following questions:

1 Does it adhere well to the wall?
If not, consider:
- Whether your backing was thoroughly dampened immediately before applying the plaster.

- Adding a bit more clay to encourage more suction between the cob and the earth plaster/render.

2 Has the plaster or render cracked excessively? If the answer is yes, consider:
- Adding more sand and straw to the mix.
- Decreasing the amount of water added to mix before applying. More water encourages more shrinking, and hence more shrinkage cracks.
- A smooth trowelled finish will create more cracking than a rougher, hand-applied finish because it creates less available surface area to release moisture.
- A small amount of cracking can be minimised and eliminated through re-trowelling or sponging when the plaster is 'green hard'.
- Try using coarser sand to decrease shrinkage cracks.
- Consider whether you applied the mix too thick. Do not exceed 12mm ($1/2$") for the scratch coat, and 6mm ($1/4$") for the final coat.

3 Does the plaster dust off when touched or when rubbed against? If yes, consider:
- Your mix may be too high in sand, and will need more clay to be added to improve its strength and cohesiveness.
- Try adding a binding agent such as cow dung or wheat paste (described below) to strengthen the mix and make it more resistant to wear and tear.

Carole Crews, of Gourmet Adobe, New Mexico, USA, encourages the addition of spoiled milk, yoghurt, mashed potatoes and even the blending up of noodles to strengthen the earth plaster mix – up to ten per cent proportion – it's all about experimentation!

Binding agents

If your subsoil is not very clayey or sticky, and is high in sand and silt, you may need to add a binding agent to improve the overall strength, adhesiveness and workability of your earthen plaster. There are many options, but the two we most favour are wheat flour paste and animal manure.

Wheat flour paste

To make a wheat flour paste, all you need is cheap wheat flour, water and to follow the instructions below.

1 Whisk together one cup of wheat flour and 2 cups cold water into a smooth paste.

2 Bring 6 cups of water to a rolling boil in a pan.

3 Slowly add the flour and water paste to the boiling water stirring constantly.

4 Keep stirring, maintaining the boil, until the paste becomes thick and translucent. Your paste is now ready to use.

If your paste remains lumpy, simply put it through a sieve. Make only enough to last for one day, although it can be kept in a refrigerator for up to one week. A general guideline is half a cup wheat flour paste to a five-gallon bucket of earthen plaster. However, it is a good idea to include this within your test batches to see how it performs.

If you are on a work site, with no access to a cooker, we have successfully used a camping stove to make the wheat flour paste.

Animal manure

Cow dung contains enzymes, which gives it inherently good binding properties. These can help to make an earthen plaster stickier and therefore more workable, and will also increase the durability and water resistance of the finished product. Horse manure can also be used, but contains less enzymes than cow dung. However, it does have the added benefit of a higher content of perfectly sized fibres which will strengthen the mix and reduce cracking, so a mixture of the two is ideal. Whichever you decide to use, always use it as fresh as possible, soaking for a short while in a small amount of water to break it down into a smooth paste. However, cow dung is normally the perfect consistency straight from the field, if you have this luxury! To determine how much to add really needs some experimentation – start off by adding about a quarter of a bucketful to every one bucket of earthen plaster. Believe us when we say that it is a truly wonderful experience to use a plaster with fresh manure in it.

Tools and materials

The tools needed for mixing and applying earthen plasters are simple and low-tech.

You will need exactly the same selection as for mixing and applying lime renders and plasters with the addition of 12mm (^{1}/2") or 3mm (^{1}/8") fine mesh screen for sieving clay.

Sieved clay

This can be the same stuff you have dug up for your cob walls. If it is from a different source, you will need to carry out the same tests described in Chapter 3: *Identifying & testing soils,* page 45, to determine the suitability of the subsoil for making a good earthen plaster. Exactly the same rules apply: a good mix is roughly 15%-25% clay and 75%-85% silt and sand. If your subsoil is too clay-rich, you will need to add more sand; if it is too sand/silt-rich, you will need to add an outside source of clay, or you can add a binder such as animal manure or wheat flour paste as described on page 153. Your extracted clay will need to be sieved through a 3mm (^{1}/8") mesh screen to remove any large stones or debris that will prevent you from achieving a smooth creamy plaster.

Sand

If the addition of sand is necessary to make your earthen plaster, you will need a well-graded, clean sand. Coarse sand can be used in the scratch coat, and a finer one if you decide to do an earthen plaster final coat.

Fibre

The fibre is very important, as it provides a reinforcing meshwork that holds the mix together. It also allows for flexibility in the plaster and reduces cracking. Traditionally in the UK, animal hair was used – cow or horse – but we prefer to use chopped straw: 100–200mm (4"–8") long in a scratch coat, and 25–50mm (1"–2") long in a final coat. We find that straw clumps less than hair during application, and in a final coat it looks very attractive, especially when the plaster has been polished or sponged to expose the golden flecks. The straw must be extremely fresh, or have been stored well, with no evidence of mould. The straw can be chopped with a weed-strimmer by placing loose strands into a dustbin (half-filled).

Left: Earth rendering an exterior wall. Middle: A good earth plaster mix should resemble a stiff cake batter.
Right: A freshly applied earthen plaster revealing the golden flakes of straw.

Alternatively, you can use a fine wire mesh screen to grate the straw through, simply by rubbing handfuls along the wire surface and than collecting the short bits in a barrow beneath. Wearing gloves and a mask is a good idea for this process.

An exciting option that we recently discovered was a product called Medi-bed which is a bedding material for horses. It comes in perfectly sized strands and is guaranteed to be dry, fresh and mould-free. We have also experimented with using Hem-core – another horse bedding material – composed of short strands of hemp, which makes for a deliciously rich earthen plaster finish. You will also need a fresh supply of clean water.

Preparing the wall to be plastered

The principles here are the same as for preparing walls to be plastered/rendered with lime. Ensure that your walls are trimmed, removing any high points, and always plaster when your electrics and plumbing have been completed, to avoid having to damage completed work. If you are working on an old cob wall that has suffered some surface erosion, it is also advisable to pre-fill any large depressions, either with a cob mix or a stiffer version of the earthen plaster. (For more information on this procedure, see Chapter 12.) Walls must be thoroughly dampened with water using paint brushes or a mister immediately before commencing work. Protect all surfaces such as floors, windowsills and woodwork, although earthen plasters are much easier to clean up than lime.

Mixing earthen plasters and renders

There are a variety of options here:

1 For small jobs, you can simply use a shovel and a sturdy wheelbarrow, turning and cutting the sand and earth as you slowly add water, and then adding fibres when the sand, clay subsoil and water have been thoroughly mixed. The fibre will absorb some of the water, so more may be needed to achieve the desired consistency.

2 Earthen plasters can also be mixed just like cob, with your feet on a large tarp (see Chapter 4: *How to make a cob mix,* page 57). This is a perfect opportunity to feel the earth against your bare feet, because unlike cob, the earthen plaster consists of sieved clay, so there will be no sharp stones to cut your feet. Similarly to making cob, the dry ingredients will need to be mixed together first, before adding water, stomping and turning. The fibre should be added at the end and thoroughly mixed in.

With all the above methods, the golden rule is to be careful about how much water you add. If at any point the mix gets too sloppy, add more of your dry ingredients in the right proportions. If adding a binder, such as wheat flour or animal dung, add at the end of the mixing process before adding the fibre.

A good workable mix should resemble a stiff cake batter. A good test is to throw a handful of earth plaster onto the wall from several yards away. It should stick well to the wall and hold its form.

Note the following, for best results:

- As with lime plastering or rendering, never apply in direct sunlight.
- Avoid working in heavy rain or freezing weather.
- Protect plastered walls with tarps or hessian for at least a week after application if bad weather is imminent.

Application

The true benefit of working with an earthen plaster as opposed to lime is that you can really experience the feel of the material. It is best applied by hand. For a scratch coat it is perfectly acceptable, and even preferable, to leave a rough hand-applied surface to provide a key for the next coat. If doing a final coat, a smooth finish can be achieved simply by hard trowelling the hand-applied material. It can either be thrown on and worked in, or smeared on using a cupped hand. For scratch coats, up to 12mm ($1/2$") thickness is acceptable, and for a finish coat, 3–6mm ($1/8$"–$1/4$") is ideal. To achieve a beautiful finish on your final coat, wait until it is 'green hard', and then either hard trowel it with a steel trowel or buff it with a damp sponge in broad circular motions, to remove any irregularities or trowel marks, and to reveal the beauty of the golden straw.

Natural paints

Naturally, any finish applied to a raw cob wall or plastered/rendered wall must follow the principle of breathability. That is, it must be porous, and have the ability to allow a healthy exchange of moisture in and out of the wall. Any latex or oil-based paints will not work. At worst, they may jeopardise the cob wall beneath by trapping in moisture, at best you will find that they will peel and flake off after a short while.

Natural paints and finishes serve the purpose of protecting the plaster or render and the cob wall, being the first line in defence against the elements and general wear and tear. They also add incredible beauty to the finished wall, and with the addition of natural pigments, the suggestions listed below will all produce wonderfully soft matte surfaces, helping to accentuate the rounded curves and forms that cob walls naturally make.

We enjoy working with a range of natural paints and finishes, most of which can be easily made at home with inexpensive ingredients. Lime wash and casein (milk) paints have traditionally been used in the UK and throughout Europe for centuries. Others we have mentioned such as an alis paint (see page 159) and litema (see page 163) are drawn from further afield: the former, alis, comes from the adobe tradition of the southwest USA, and the latter, litema, from the earth building traditions of South Africa and India. Our final suggestion: natural, commercially produced paints made solely from plant oils, and hence completely breathable, are a good option for those wanting a more conventional and homogeneous finish.

Natural pigments

Home-made paints can all be enhanced and literally brought to life with the addition of natural pigments. We use only naturally occurring mineral and earth pigments, which we source from different suppliers (see Resource list, page 241) or obtain from different clays dug straight from the ground, which can be ground or crushed into a fine powder using a pestle and mortar. Natural pigments from organic and inorganic substances, such as minerals, rocks and the earth, have been used since pre-historic times. They provide a variety of muted hues and tones, the most familiar being the red and yellow ochres and green earths, which are derived from the iron found in the Earth's crust.

Lime wash

Lime wash is a very simple, inexpensive and effective finish made from either lime putty or Natural Hydraulic Lime (NHL 2 – in powder form) and water, with the additional option of pigments. A well-made lime wash will carbonate on the wall, changing into calcium carbonate, and produce a durable, water-resistant finish that will not dust off. For the beginner plasterer, applying a lime wash has the added benefit of masking any small cracks that may have appeared in a final coat of lime plaster/render. The converted calcium carbonate will fill the cracks.

To make a simple lime wash from lime putty:

1 Use a large bucket: add two parts water to one part good quality lime putty, which has been pre-beaten (like you would a cake mix) to remove stiffness.

2 Mix vigorously, either using a hand whisk or a drill with paddle attached, until it reaches the consistency of milk.

3 Pass this milky mixture through a sieve to remove any lumps, gritty particles etc. If adding pigments . . .

Left: Adding pigment to a bucket of lime wash. Middle: Lime wash and pigment should be mixed thoroughly with a paddle attached to a drill (which is commercially available). Right: Applying the lime wash with a thick-bristled brush.

4 . . . add the desired amount of pigment to a jar of warm water (it will also work with cold) and shake vigorously (with the lid on tightly!) until thoroughly mixed.

5 Add the pigment and water solution to the bucket of lime wash and mix together with whisk or drill and paddle.

Making lime wash from NHL 2

1 Add half a part of powdered lime to one part of clean water, and use a whisk attached to a drill to mix thoroughly until a consistency of milk is achieved.

2 Add pigments if desired, as described above.

A note about adding pigment to lime wash
The colour of your lime wash with pigment will dry significantly lighter than the colour of the raw pigment, and how it appears once mixed up in the bucket. To know how much pigment to add will take experimentation and a series of test patches, allowed to dry fully (at least one day) on the wall to be lime washed. Start off with a few table-spoons of pigment, and increase or decrease accordingly. Don't feel restricted to using one colour: you can mix many different colours to produce some unique shades.

Application

1 Before applying, ensure the wall is free of dust and thoroughly dampen it down with a fine mister immediately before application. Protect floor surfaces and wood detailing, as lime washing can be a messy business. Use thick-bristled brushes 100–150mm (4"–6") wide, and smaller brushes for detail work around windows, doors and edges.

2 Keeping a well-coated brush, apply the lime wash in broad circular strokes, ensuring all drips down the wall are worked in well.

3 The lime wash will appear translucent as it goes on, but will soon turn opaque as it carbonates and dries. In the initial coat, it may be hard to see where you have worked, but any missed patches can be filled in on subsequent coats. A freshly lime-plastered wall will need at least three coats to achieve a deep colour.

Tips for best lime wash results
As you progress along the wall, or if you take a break and the bucket of lime wash is left standing, give it a good mix with the drill and paddle. A stir with a trowel will also suffice. This is to prevent the pigment and lime from settling on the bottom of the bucket. Failure to do this will result in an uneven distribution of colour.

Try to avoid lime washing in direct sunlight, as it will dry out too quickly and hence not fully carbonate, manifesting in a dusty finish. Cover with hessian or damp sheets to protect from the sun for at least 24 hours after application. Avoid lime washing in the heavy rain. The best weather is a damp, cloudy day.

At least this weather is good for something!

Always wait at least 24 hours before applying your next coat of lime wash, and mist down the wall surface lightly before the application of each new coat. The rate at which the lime wash dries will depend on the wall surface beneath it. If your wall has a mixture of different substrates such as stone or cob, or different parts of the wall are more damp than others, the lime wash will initially appear mottled but will even out once all areas have dried out fully.

Do not make the lime wash too thick, as it will crack and craze excessively on the wall as it dries out.

Alis

At its simplest, an alis is a thin clay slip that can be applied to internal and external earthen walls that have been finished with an earthen plaster, for protective and decorative purposes. As a lime wash finish is the perfect partner to a lime plastered or rendered cob wall, an alis is the perfect partner to an earthen plastered or rendered cob wall (although lime wash can be used).

They come predominantly from the adobe tradition in the south-west USA, and were traditionally applied by the Native American women who were known as the 'enjarra-dorras' meaning 'plasterers'. The alis was made using local clays and applied with sheepskin. As natural builders today, constructing new earthen structures, we can tap into this functional and creative wisdom.

Like a lime wash, an alis will allow the cob wall to breathe, but will produce a more lustrous surface as opposed to the flatter matte finish of a lime wash.

As mentioned in the earthen plastering section, in a British climate an alis is best suited for internal use or on very well sheltered external walls. They will not stand up well to direct exposure to the rain.

An alis can be made from a basic set of ingredients comprising simply clay and water. It can be made more durable with the addition of flour paste, and more decorative with the addition of mica (a naturally occurring mineral, which comes in flakes or powder form and will increase the durability of the finish as well as producing a lustrous sheen when buffed), chopped straw and pigments. You can simply use the same clay subsoil that you used to create your earthen plaster, or seek out some interesting coloured clays naturally occurring in the ground around you, or from other local building sites. This will enable you to create a unique palette of colours for your walls.

If you want to produce a pigmented alis finish, you can purchase and use white kaolin clay from a pottery store, which will give you a neutral starting point from which you can then create any colour you want.

To make an alis clay paint

1 Screen clay subsoil through a 3mm (^{1}/8 ") wire mesh screen to remove debris and

" ... natural finishes come in all shades, hues and colours. Always individual, subtly varied and rich, they are simply a joy to the senses ... "

large stones and grit. Kaolin does not need to be screened as it already comes in a fine powder form.

2 Make a fresh batch of wheat flour paste (see page 153).

3 In a large bucket or tub, add together: 1 part flour paste; 2 parts kaolin or sieved clay; 2 parts fine silica sand or mica (or a combination of the two).

4 Mix together with a whisk or paddle attached to a drill to create a lumpless, smooth mixture resembling thick cream. Add more water if necessary.

If adding a pigment:

5 Mix pigment with warm water in a jar and then add to the alis mix in bucket. (Be sure to account for the extra water in your pigment.) Mix alis with pigments.

6 (Optional) Add a few large handfuls of chopped straw (see earthen plasters section, page 150) and mix evenly.

Always do a few test patches before you begin, then ask the following questions:

- Is the finish smooth or grainy?
  If grainy, you have added too much sand, or the sand is too coarse.
- Does the finish dust off easily? If yes, you have added too much sand or may need to add more flour paste or clay.
- Does the alis evenly cover the wall? If no, you have added to much water.

Application and polishing

1 Prepare the walls before applying the alis finish. Ensure they are dust-free, and lightly dampen the surface using the fine mist sprayer. Protect all floor surfaces and the wood around windows and doors.

2 Apply the alis using a large paint brush (you can try sheepskin if you like!), ensuring all areas are evenly covered.

3 Allow the first coat to dry for a day, and then apply a second coat. Two coats are usually sufficient.

4 When the second coat has set and is still moist, polish using a damp clean sponge to remove excess dust and expose the shiny flakes of sand, mica and/or the golden straw. Rub in circular motions, rinsing the sponge if sand particles build up on it.

An alis can easily be reapplied yearly, or when you want to freshen or change the colour of your space.

Casein paints

Casein paints are also traditional home-made paints, which have been used throughout the world for thousands of years, but which fell out of fashion with the introduction of modern paints. They are also sometimes referred to as milk paints because the casein comes from a protein found in milk. It is possible to make a casein paint starting right from scratch, separating the curds and whey from milk, but casein powder can also be bought fairly inexpensively (see *Suppliers* section, page 241), and this considerably simplifies the process.

On its own, a simple mixture of casein and water applied to an earthen plaster finish will provide a translucent glaze allowing the earthen plaster to shine through beneath. This will have the added benefit of minimising the dusting that can sometimes occur in an earthen plaster. To make a full-bodied, opaque paint finish with or without pigments, a filler can be added, such as clay, chalk or lime, to produce a matte finish similar to a lime wash.

They are suitable only for an internal finish and are water-resistant but not waterproof. Don't use as an external finish. We use a recipe from the American straw-bale builders Bill and Athena Steen.

To make a simple casein paint with casein powder, you will need:

- Casein powder
- Water
- Borax powder
- A filler of chalk, clay or lime
- Optional pigment.

Method

1 Make a slurry of casein and water by adding together 2 parts casein to 8 parts water.

2 Allow this mixture to sit overnight or for at least 2 hours.

3 Add together one part borax powder purchased from a chemist (or other listed supplier in the *Resources* section, page 241) with two parts warm water.

4 Add together the borax powder and water solution and the casein slurry, and to this add another 6 parts water.

5 For every one part of the borax and casein solution, add nine parts of a solid filler (chalk, lime, clay), if desiring an opaque paint.

6 If you desire a colour for your casein paint, add desired pigment powder to a small amount of the casein and borax solution, mixing well.

Add this to the casein and filler combination. You can also use coloured clay as a filler, which will also act as a pigment to your finish.

Application
As with all home-made paints, you will need to run a series of test batches to ensure that the proportions are right for your ingredients and wall substrate. Apply with a large paintbrush, misting the wall down as you go.

• If the mixture dusts when rubbed, you need to add more casein powder.
• If it cracks and peels, the mixture needs to be diluted with more water.

Litema
This is a traditional clay dung plaster finish, extensively used in Africa and India, where

external decoration of buildings is a fundamental part of cultural expression. Litema utilises two very simple ingredients, available everywhere: clay and animal dung. While training in Oregon, we learnt how to mix and utilise these materials into decorative relief around doors and windows, which also affords protection for these vulnerable areas. We can use the materials to the same effect as our brothers and sisters in Africa and India, and just as beautifully! However, we have learnt never to apply litema in damp, cold weather, as it will take ages to dry and may mould, and unfortunately it will not stand up to intense British horizontal rain. So only use it in less exposed or protected areas such as porches or covered outdoor areas.

Method

1 Collect either fresh or dry cow or horse dung (cow dung is best).

2 Mix with equal parts of wet clay subsoil.

3 Allow this mixture to ferment for a couple of days, and after trying a small test patch on the wall to be 'litema'd', simply smear on to a pre-dampened cob wall with your hands. Gloves can be used by the faint-hearted, but it is entirely safe to go barehanded – just wash your hands afterwards! Shape the relief as you desire, building up in several thin coats. If your litema cracks, you simply need to add more manure to the mix.

Manufactured breathable paints

If all this talk about home-brewed natural paints is a bit overwhelming, there is another option: you can purchase some excellent natural, fully breathable paints made out of pure plant oils off the shelf from specialist suppliers (see *Resources*, page 241). These will remove any of the guesswork and experimentation needed for the above examples and will produce consistent colours and finishes. Be warned: some companies such as Dulux have started to bring out some so-called 'breathable' paints. However they are not fully breathable, and are relatively toxic as compared to natural paints. Stick with reputable suppliers who sell products made by companies such as Aglaia natural paints and Beeck'sche Farbwerke from Germany. They are expensive, but are reliable, non-toxic, and beautiful to use and look at on a final coat of plaster. They smell divine, and will leave a room freshly scented for weeks to months after their application. These are best used directly onto lime plaster, needing only one or two coats. Do not use them directly onto cob or earthen plasters, as the clay will interfere with the finish and give a muddied look to the colour. We normally use these natural paints to good results in restoration projects to tie in walls with different substrates, such as a lime plastered wall next to a cement plastered wall in the same room which will not take a lime wash. Aglaia natural paints do a historic colour range to blend with the colours of traditional buildings.

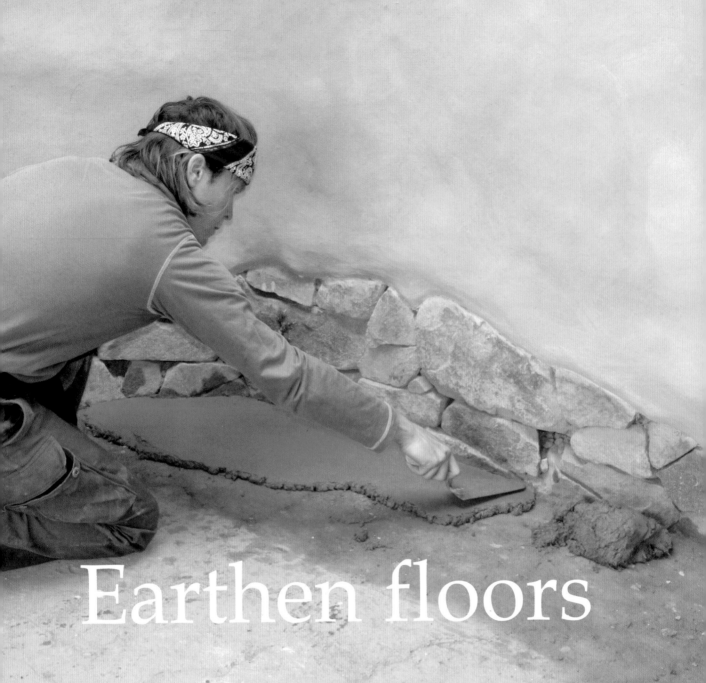

Earthen floors

10

. . . beautiful, non-toxic, extremely durable,
easily repairable, inexpensive, warm.
Some of the endearing properties
of a natural earth floor.

Chapter 10

The floor is the area in your house with which you have the most daily physical contact. By creating an earthen floor, you will be re-shaping your perception of what the function of a floor is. You will no longer see it as just something to walk on, and keep the cold and damp of the ground below out. An earthen floor is more than these: it is both functional and beautiful.

You will want to lie on it, touch it, look at it, and walk barefoot on it. Earthen floors are beautiful, non-toxic, extremely durable, easily repairable, inexpensive, warm, kind to the planet, and can be tailor-made with local, coloured clays. An earthen floor consists simply of layers of compacted earth (similar in mix to cob), laid on top of a free-draining sub-grade layer, and finished with linseed oil and beeswax to create a durable, shiny finish.

Historically, the concept of an earthen floor grew out of the earliest houses, whose floors were made from the natural earth of the ground on which they were erected. At their simplest, floors were levelled, swept, sprinkled with earth and tamped. At the more elaborate end, bullock's blood, fine clay and bone chips were added to produce an almost black marble finish.

The following recipe and technique is an example of an mid-seventeenth century English earthen floor, described by one Henry Best: "The earth was to be dug and raked until the moulds were 'indifferent

small'. The water was to be brought in 'seas' and 'great tubs or hogsheads or sleddes'. The earth was then to be watered until it was a 'soft puddle'. It was then allowed to lie a fortnight until the water had settled and the material had begun to grow hard again. Then the floor was to be 'melled' and beaten down with wooden paddles to a smooth finish." (Robert Edmunds, *Your Country Cottage).*

The earthen floor that we are going to describe has been adapted from both the European earth building traditions, as described in the quotation above, and the North American adobe tradition. It takes into account the often wet British climate, and the need for protection from the cold ground. It therefore allows for extra insulation and drainage.

In Chapter 2 (page 33) we talked about the concept of passive solar design, and how materials with good thermal mass, such as cob, are needed to maximise this, through their ability to absorb and then release the heat from the sun. The floor is perhaps the most important area for this heat absorption, as it receives more direct sunlight than the inside of the walls. It is important to utilise this, and therefore ensure that the floor is made of a dense material that has a high thermal mass rating. Earthen floors are an excellent way to embrace this passive solar concept and welcome nature's free heat from the sun into our houses. Compare this with a modern-day cement floor, which has a high thermal conductivity and will therefore constantly draw the heat from your feet into

tools

1	Shovel for digging, mixing, etc	(4"x2") battens.
2	Tamper – a home-made piece of round wood with good grip grooves, or a commercially available square-ended tamper	5 Trowels – floats and gauging trowels
		6 Sponges
		7 Stove to heat beeswax
3	Pitchforks	8 Paint-brushes for applying linseed and thinners
4	Levels – long ones attached to 100x50mm	9 Lint-free rags to apply beeswax

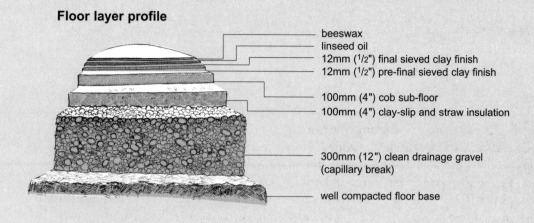

Floor layer profile

- beeswax
- linseed oil
- 12mm (¹/2") final sieved clay finish
- 12mm (¹/2") pre-final sieved clay finish
- 100mm (4") cob sub-floor
- 100mm (4") clay-slip and straw insulation
- 300mm (12") clean drainage gravel (capillary break)
- well compacted floor base

the ground, making it extremely cold and uncomfortable to walk barefoot on.

The floor consists of a simple layering system, starting from a solid well-compacted sub-floor, on top of which is installed a 300mm (12") layer of gravel, 20–40mm (³/4"–1¹/2") in size). On top of this is laid a 100mm (4") clay-slip and straw insulation layer, and then a 100mm (4") sub-floor, made out of a standard cob mix with the addition of extra aggregate.

The penultimate layer consists of a 12mm (¹/2") sieved clay/fine sand/chopped straw mix, hand-trowelled on, and the final layer consists of a 12mm (¹/2") smooth trowelled mix, the same as the one above it. The floor is finished with four layers of boiled linseed oil in varying proportions, on top of which a final layer of beeswax and linseed oil is rubbed in.

Materials and tools

The necessary materials for making a good earthen floor are very similar to those making a cob wall mix. Again, they are simple, inexpensive, easy to obtain, and for the novice earth builder, easy to use.

To make an earthen floor you will need:

- clean gravel for base layer.
- A clay rich sub-soil for insulation layer, sub floor and final cob mix.
- Clean, long straw for insulation and sub-floor cob mix, and short, chopped straw for smooth finish cob mix.
- Well-graded coarse sand for making sub-floor cob mix.
- Boiled linseed oil and turpentine or citrus thinners for hardening and sealing.
- Beeswax for final polishing and waterproofing.

Test batches

As with all earthen building techniques that involve the use of non-standardised, local materials, you will need to carry out a few test batches of your flooring mixes to monitor for hardness, durability, shrinkage and cracking.

Test the subsoils to be used, as described in Chapter 3: *Identifying & testing soils*, page 45, to ensure that the earth you are going to use is suitable. For the sub-floor, make up a few cob mixes as described in Chapter 4: *How to make a cob mix*. It is necessary to use more sand in the floor mix to prevent the likelihood of cracking, and to add some of the gravel used in the base layer to help

Laying an earthen floor

Dig down to solid ground and level off.

Install layer of gravel – at least 300mm (12") deep – and tamp well.

Insulation layer of straw and clay slip – at least 100mm (4") thick. Compress well.

Sub-floor of 50–80mm (2"–3") of cob mix. Make sure you keep your levels, and tamp well.

Final layer of fine sieved clay, sand and chopped straw. Apply with trowel or float.

Ready to seal and waterproof floor. Application of linseed oil with brush and beeswax with cloth.

An earthen floor should only be installed in a building that is very well drained. The finished floor level should always be at least 150mm (6") above outside ground level height.

Steps 1–4 can be done before the roof is on and the walls are fully built. The foot traffic will help compact the sub-floor layer, but ensure you protect the floor from prolonged and frequent rainfall.

If you are installing an underfloor heating system, the pipes should be buried within the sub-floor layer.

The linseed oil should be mixed with a thinner for maximum penetration, and at least four coats applied.

The melted beeswax should always be mixed with linseed oil, to create a paste. Apply when cool.

strengthen it. For a finish layer mix, the clay needs to be sieved through a 3mm ($^1/_8$") wire mesh to remove stones, debris, and finer sand added. Chopped straw, 'Medibed' – a horse bedding which can be purchased from an agricultural store – or regular straw chopped with a weed whacker (see Chapter 9: *Lime & other natural finishes*) should be added instead of long straw. Create samples 25mm (1") thick and 300mm (12") square, laid onto a flat base out of the rain. Allow the samples to dry thoroughly, and monitor for cracking and strength. Remember to take a note of your mixes.

If your sample cracks excessively, add more sand and straw, and ensure the mix is made as dry as possible. If your sample is brittle and not very durable, add more clay to improve its strength.

Preparing and laying the earthen floor

It is vital that the site on which you plan to create your earthen floor is well draining – this should have been considered in the siting of the house at the initial stages of planning and design. The final layer of the floor must always be laid at least 150mm (6") above the outside ground level of the building to prevent the possibility of the incursion of moisture ruining the floor.

1 Prepare the ground. The ground for creating an earthen floor can be dug out and prepared at the same time as your foundations are dug. Create a solid platform, removing all organic matter and topsoil.

The ground must be well compacted with a tamper: home-made with a long piece of round wood (see middle photo on page 166), or a standard square tamper purchased at a builders'-merchant. For very large spaces, a pneumatic tamper can be used. When digging and creating your foundations, plan ahead to accommodate for the total depth of the final floor. To do this, add up all the different layer heights and mark off on your foundation plinth. For most parts of the UK, a good earthen floor should be around 500mm (1' 8") in total depth, including the gravel base and insulation layer.

We do not recommend the use of an impermeable, plastic damp-proof course under an earthen floor. This is because any impermeable layer may cause condensation to form, possibly leading to moisture entrapment and damp problems. The gravel base layer will allow moisture to drain freely and hence prevent any moisture from the ground rising into the walls. However, this should not happen if you have sited the building appropriately and hence the floor.

2 Install your gravel base layer, made of 300–360mm (12"–14") of clean gravel, which must be thoroughly compacted using a tamper (see Step 2, page 168). This must also be levelled using a long level attached to a 100x50mm (4"x2") batten, which can be used to scrape off any excess material. It really helps at this stage to mark out the height of each level onto your foundation plinth, so that you can see exactly what height each layer will come to.

work towards door

check level

screed board

fill channels with flooring mix

Laying the sub-floor

Place squared-off timber guides on your compacted gravel layer, and check for level in all directions. Next fill the bays between the timbers with the sub-floor cob mix, tamp down and level off with a screed board. Take out timber guides once the bays are filled, and also fill with flooring mix. Do this a bay at a time, working your way towards the exit door.

3 Install an insulation layer to prevent heat loss from the building leaking into the ground, and the cold from the ground seeping into the building. There are a few options here: the least favourable option is the standard rigid foam insulation such as Kingspan, which is environmentally destructive but will provide excellent insulation results and will last for a very long time. Alternatively, a 100–150mm (4"–6") layer of clay slip (a soupy liquid made out of sieved clay mixed with water) mixed with straw can be used. A more expensive but less labour-intensive method is to use perlite or vermiculite bound together with a wet clay subsoil and laid in a 100–150mm (4"–6") thick layer. This mixture can be trodden and mixed on the ground or on a tarp the same way that you would make a cob mix.

4 Ensure that the insulation layer is thoroughly dried before laying the sub-floor layer. This consists of a similar mix to the cob used to build the walls, but should contain more sand and some gravel. This layer provides the main mass of the floor. It should be made as dry as possible to minimise cracking, but needs to be wet enough so that it spreads on smoothly and evenly.

Installing the sub-floor
(see diagram above)

• Lightly dampen the insulation layer. This is to key the next layer to the base.

• Insert two lengths of 100 x 50mm (4 "x 2") wood, standing on their edges, 900mm (3') apart and spanning the full width of the space. Ensure that they are level with each other and at the correct height.

170

- Using a pitchfork or your hands, dump the cob mix onto the floor area in between the two boards.

- Spread the cob out evenly and tread into the layers below, ensuring it is well compacted. Small thwackers and mallets can also be used to tamp it down. The cob mix must come to the top of the lengths of the 100 x 50mm (4"x 2") pieces of wood.

- Using another similarly sized piece of wood as a screed board, lay on top of the two guides and scrape off any excess cob.

- Take away the guide board farthest away from you and immediately fill the gap, ensuring the cob is well bonded.

- Place the removed guideboard three feet in front of the remaining guide board and repeat the above process until the floor is covered.

It is necessary to create a fairly rough finished surface in the cob to create a good key for the next layer.

Allow the sub-floor to dry thoroughly before walking/working on it. Good air circulation will hasten this process.

5 Apply the penultimate 12–25mm ($^{1}/2$"–1") finish layer. Before you lay the final layers, sweep the sub-floor to remove dirt and dust. Dampen the sub-floor to ensure a good adhesion with the final layers. The final layer mixes are made with sieved clay subsoil through a 3mm ($^{1}/8$") screen. Finer sand is used, and chopped straw instead of long straw. This mix can be made on a tarp in exactly the same way as any other cob mix. Use a very high sand content (refer back to your test batches to find out how much), as it is important to minimise the cracking in these layers. This penultimate layer should be no thicker than 25mm (1"), and should be trowelled on in very much the same way, as you would plaster a wall. You can tread it in with your feet to ensure good adhesion with the sub-floor below.

A rounded pool float or gauging trowel can be used. To keep a good level, use a level taped to a straight board 900–1200mm (3'–4') long, which can also be used to scrape off any excess material.

Again, the finish should be level but not completely smooth, to provide a good key for the final layer.

6 Apply the final layer when the previous layer is 'green-hard'– hard but not completely dry. If the previous layer has cracked, you will know to add some more sand to your mix. Apply the final layer 12mm ($^{1}/2$") thick, in exactly the same way as the previous one.

If you have a source of special coloured clays, use these in this final layer, as the final oiling and waxing will maintain the exact colour of the wet material. When the surface is 'green-hard' it can be misted lightly and hard trowelled to polish smooth, or buffed with a damp sponge to expose the beautiful flecks of straw – an exciting moment!

7 Apply your linseed oil sealant. Keep all foot traffic off the floor until it is thoroughly dry. At this point the floor is ready to be sealed to make it waterproof and durable, and to create a rich sheen.

This is a good time to fill in any cracks that may have emerged. Simply dampen the concerned area and use a sponge to work some of the final floor mix into the cracks.

In the past, animal urine and blood were used to achieve the above. These days it is more common to use boiled (not raw) linseed oil mixed with a solvent such as turpentine, or citrus oils for a more environmentally friendly option (see *Resources*, page 241).

The solvent acts to assist penetration of the oil deep into the pore spaces of the floor. Without it, the oil will simply form a shell on the surface of the floor, which will be easily broken and damaged.

The formula is as follows:
1st coat: 100% boiled linseed oil
2nd coat: 75% boiled linseed oil/25% thinner
3rd coat: 50% boiled linseed oil/50% thinner
4th coat: 25% boiled linseed oil/75% thinner

- Mix the oil and the solvent together in a bucket, and apply evenly using a paintbrush.

- Allow each coat to dry between each application, and when the floor no longer feels tacky to the touch.

- Ventilate the room well to assist the drying process.

The oil can be warmed to assist penetration, but this is not necessary.

8 Apply final beeswax polish. Allow the final coat of linseed and solvent to dry before applying the ultimate and most revealing of the finishes to the floor – the beeswax polish. This makes the floor waterproof (it will be possible to mop and sweep) and very shiny, bringing out the flecks of straw from the final earth coat.

You will need a bar or granules of hard, raw beeswax, which should be melted in a bowl, in a pan of boiling water on a stove. Allow the beeswax to become completely liquid, and while still warm, add some boiled linseed oil in a ratio of one part wax to two parts linseed oil. The linseed oil must be added to enable the beeswax to form into a paste. Without it, the beeswax will simply harden the moment it cools down, making it impossible to spread on the floor. Allow this mixture to form into a paste, which is then spread onto a clean floor using a lint-free cloth. Use a circular, polishing motion, really working the wax into the floor.

Keep off the floor until thoroughly dry. To test the effectiveness of the beeswax, pour a small amount of water onto the floor and watch it bead up.

Top left: Placing straw-rich clay slip insulation layer.

Top middle: Levelling the sub-floor.

Top right: Wetting the sub-floor to key the final layer.

Above: Treading the final layer to achieve a good adhesion to the sub-floor below.

Above middle: A section of the final layer floated onto the subfloor.

Right: Impregnating the floor surface with thinned-down linseed oil.

Above: Raw granules of beeswax before being melted for final polish and waterproofing.

Left: Polishing the beeswax into the floor.

Right: Finished floor polished to reveal the golden flecks of chopped straw.

Right and below: Ceramic floor tiles used as an alternative to earth floor in a bathroom or kitchen.

Far right: Thick cob floors play an essential role in storing heat as they have excellent thermal mass.

Far right: Example of a pneumatically compacted floor, oiled and wax-finished in a commercial warehouse.

Maintenance and repair

The oiling and waxing procedures will make an earthen floor extremely durable, and able to withstand most of the knocks of daily living. Maintenance simply calls for a yearly beeswaxing session (or more frequently if you wish), just as you would polish a wooden floor. This will maintain the beautiful sheen and help protect the earth underneath.

If the floor does get damaged, simply break away the loose portion and build up again with the original (or a freshly-made) mix. If it is simply chipped, the flooring material can be rubbed in with a sponge. Oil and wax the whole floor to blend the new repair into the existing floor.

efficient material to use as a heating system in this way. It makes more sense to use earth with under-floor heating than wood, because wood is very efficient at holding heat in (i.e. it is a good insulator), which will lead to most of the heat produced from the hot pipes being trapped underneath the wood, and therefore unable to radiate throughout the house. The added warmth of the under-floor heating in the earthen floor adds to the pleasure of moving and living on top of the earthen floor.

The pipes should be fully buried within the middle of the sub-floor layer, the earth replacing the cement in a conventional system. It is important to test for potential leaks in the pipe system before covering with earth.

> Floors receive more direct sunlight than walls; so an earthen floor with its inherently good thermal mass will serve as an excellent heat store.

It is a good idea to make some extra floor mix to keep aside for future repairs. It can be dried out and simply re-moistened when needed.

Under-floor heating

Earthen floors are very compatible with radiant under-floor heating. The ability of cob to hold and then release heat makes it a very

Other natural flooring options

In areas where there may be excess amounts of water sloshing around, such as bathrooms, it would be wise to use other flooring options, such as slate or ceramic tiles.

Fireplaces

A carefully designed and well operated earthen fireplace can provide heat and ambience in a modern, ecological home.

Chapter 11

We talk a lot in this book about the satisfaction and practicalities of building with the basic elements of mud, stone, grass and wood. To make good cob it is necessary for the elements of water and air to be present also – water to mix the mud, and air to dry it. The final element of fire completes cob's relationship with the elements – with a special mix, made without straw, cob is completely fireproof and therefore serves as an excellent material for containing fire.

Its ability also to store and release heat (i.e. having a high thermal mass) makes it an obvious choice for making fireplaces and bread ovens. The Iranian/American architect and visionary Nader Khalili takes the application of fire with cob one step further, and actually fires earth structures *in situ* to create ceramic houses, just as you would fire a clay pot. He claims the introduction of fire to clay creates the perfection of earth, and transforms its weakness – vulnerability to water – into a material that is completely and permanently resistant to water.

In a fireplace or earth oven context, the contact between the fire and the cob will indeed, over time, transform the material into something that resembles a low-fired brick, and though not specifically fired and glazed like Khalili's ceramic houses, will produce a material that becomes extremely hard and fire-resistant.

In the earliest shelters, structures were designed and sited to keep the cold out and to utilise the warmth from the sun as a heat source. When extra heat was needed, our ancestors burned wood on an open hearth in the middle of the room, with a smoke exit hole in the roof, and drew on the body heat created from animals and humans. When we look at nature, the methods used by ants are even more elemental than this: they heat their colonies by taking turns to sit in the sun, soaking up its radiant heat, and then return inside to re-radiate this heat back into the colony and to their fellow ants, like minute portable cob fireplaces. Consider also the wasps and the bees, who warm their hives through the heat generated by flexing their abdomens and flapping their wings. . . .

. . . But we digress. After the open hearths, the methods used evolved into fireplaces with chimneys to carry smoke away, and here begins the history of earth as a medium for making fireplaces and ovens, all over the world. To mention but a few: the ancient Egyptians made beehive-shaped clay ovens

to bake bread, and the native Americans of the pueblos made bread ovens out of adobe (sun-dried mud bricks).

More recently, in the UK, many old cob cottages dating from the 1800s contain large inglenook fireplaces, which often contained within them bread ovens in nooks carved out of the cob. In Devon and Cornwall, the traditional 'cloam' oven was made out of cob, and again came in an oval beehive shape. As will be explored later, it is no accident that indigenous peoples all over the world have designed their bread ovens in this particular beehive shape.

Ianto Evans of the Cob Cottage Company in Oregon, USA, and Kiko Denzer, also of Oregon, have done much inspirational work to combine cob with fire to heat cob buildings and create functional bread ovens. Ianto has developed the Rocket Stove, a super-efficient wood-fired heating system, which works very effectively in cob structures. He has also adapted an efficient cob fireplace, from the principles developed by Count Rumford in

17th-century cob inglenook fireplace with in-built bread and clotted cream ovens.

the late 1700s, and has adapted traditional earthen oven concepts into a simple, accessible design that can be easily created in your back garden to produce endless batches of yummy home-baked pizza and bread.

There is not enough room in this book to talk about the rocket stove, but we will describe in detail how to make a cob Rumford fireplace and a cob bread oven. To learn more about Rocket stoves, consult Ianto's great book: *Rocket stoves to heat cob buildings: how to build a super-efficient wood-fired heater* (see *Resources*, page 241).

The special fire cob mix

To make a cob mix that is completely fireproof, simply omit the straw (a highly combustible material), which will create a denser mix and therefore store more heat, and add a higher ratio of sand to clay (3:1 to 5:1, depending on clay content) to prevent the cob from cracking when it comes into contact with the heat of the flames. This should be mixed in exactly the same way that you would make a regular cob mix.

The place of the open fire in a modern ecological home

Open fireplaces are not considered to be an efficient way to heat an inside space, and indeed, without a proper air source, will end up sucking more cold into the house from the cracks in doors and windows than they will produce heat to warm the house. For this reason, they have lost their place in the modern, ecological home, being replaced by wood-burners and central heating systems.

They are also open to criticism because of the CO_2 produced from the burning fuel. We feel, however, that if designed and operated properly, and placed strategically within the home, the open fireplace not only can provide a good amount of heat, but has the other benefit of bringing hours of welcome relief from dreary winter weather.

An open fireplace will bring you things that no other heat source can provide. It nourishes the senses – the sight of leaping flames, the smell of burning wood, and the feel of the glow from the heat on your body; it creates an atmosphere of sharing, calmness and introspection; and it can truly provide the heart of the home, bringing together people and animals to watch the yellow flames dance and flicker into the night.

On a practical note, the following steps can be taken to improve the efficiency of an open fire:

1 Design the fireplace along the principles of the Rumford fireplace (see next section).

2 A fireplace is best placed in the centre of the house, where the cob can act as a heat store, absorbing the heat and re-radiating it. It is best not placed on an outside wall, especially the cold, north wall, where much of the heat will be lost to the outside.

3 Provide an adequate flow of air as close to the fire as possible, to prevent the fire from drawing cold air across the room from gaps in the windows and doors. In a room

with suspended wood floors, a grille can be inserted in the wood in front of the hearth; or if placed on an outside wall, an air brick or grille can be inserted behind the fireplace, with a slide mechanism to give the option of being able to close it off when the fire is not in use.

4 Always use good, dry wood to produce a cleaner burn and to use less wood – clean fires burn less fuel.

5 Always fit a damper to prevent heat from the house exiting out of the chimney when the fireplace is not in use.

A small, compact and well-designed cob house that is well insulated in the roof, foundations and floor, will benefit from the heat of an open fireplace.

The cob Rumford fireplace

Have you ever huddled around an open indoor fireplace desperately trying to get warm from the flames, but found yourself being left cold and unsatisfied? Or struggled to get a fire going at all, as you choke on the black smoke? One man – Count Rumford – an American supporter of King George III who was forced to flee the country, and reside in England in the late 1700s, dedicated his life to researching the art of the fireplace. He developed a design that would never smoke, and that projected its heat into the room and not up the chimney. His interest in the fireplace was aroused because of the ineffective designs that he discovered on his arrival in England, which consisted of ". . . massive square boxes calculated best to roast an ox and freeze an audience" (Vrest Orton, *The Forgotten Art of Building a Good Fireplace*). He set about replacing the deep and squatty designs that he so commonly found, with one that was shallow and high with slanted sides and back. Rumford started his research by exploring the nature of heat, and came to the conclusion that the heat from a fireplace was radiant heat, meaning that it is cast in straight rays from the fire into the room, and heats the bodies and objects that these rays come into contact with, and not the air. This led to a design that contained the following features:

Materials
The fireplace should be constructed out of good heat-radiating materials such as stone, or brick – and of course, though he didn't mention it, cob. The back and sides of the fire should be made out of materials that are smooth and not jagged, to aid in the projection of the heat into the room. Rumford mentions brick for this, but of course cob is perfect.

Specific dimensions
No matter what the size of the fireplace, the dimensions must follow the same rules, and have the same relative ratios. Rumford scientifically calculated these dimensions through rigorous trial and error (see page 182). These dimensions maximised the projection of the radiant rays of heat out into the room.

Left: A cob fireplace designed around the Rumford principles, designed and built by Cob in Cornwall Ltd.
Right: An outdoor cob Rumsford fireplace, designed and built by Seven Generations Natural Builders in the USA.

Bevelled sides and slanting fire-back

To help cast the heat from the fire into the room, and hence be reflected away from the back wall, the fire-back should be gradually slanting forwards, and the sides bevelled (see page 182). This also aids the passage of smoke, vapours and gases up the chimney instead of into the room.

Depth of fireplace throat

The throat (the area behind the lintel) should never be more than 100mm (4") deep, or less than 75mm (3"). If it is any larger, the necessary circulation of cold air coming down and mixing with the warm air that wants to rise, won't take place. Most 18th-century fireplaces (and indeed many modern ones) demonstrate a hugely oversized throat because enough space was needed to send young chimney sweeps from the room up into the chimney! Rumford argued that if a fireplace is designed correctly, it will burn efficiently, and the chimney will not be clogged, hence there will be no need for a chimney sweep. We have found that Rumford was right!

Construction of a smoke shelf

This is the most important feature of the Rumford fireplace. It consists of a 75–100mm (3"–4")-deep shelf (which can be constructed out of cob), which projects from the back of the smoke chamber wall 300mm (12") above the lower edge of the lintel. Its function is to mix the cold air descending from the chimney, with the hot air rising from the fire. The cold, descending air hits the smoke shelf, and bounces toward the front of the shelf, meeting the hot, ascending air. From here, the hot and cold air rise up the inside wall of the lintel, creating a suction effect to draw any smoke, vapours and gases up the chimney and out of the house. This is the ultimate step in creating a smokeless fire.

Damper

A damper should be installed to block the passage of cold air into the room from the chimney, and to inhibit the warm air from the room rising up the chimney when the fireplace is not in use. This should be placed at the entrance of the smoke chamber i.e. sitting on the smoke shelf.

Smoke chamber

This refers to the area above the smoke shelf, and in front of the flue, and is the area that

Side elevation

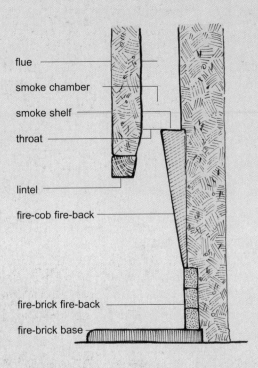

flue

smoke chamber

smoke shelf

throat

lintel

fire-cob fire-back

fire-brick fire-back

fire-brick base

Perspective

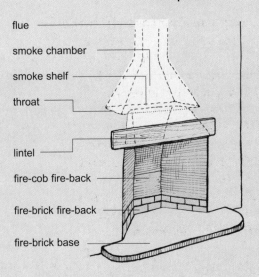

flue

smoke chamber

smoke shelf

throat

lintel

fire-cob fire-back

fire-brick fire-back

fire-brick base

facilitates the mixing of the descending cold air with the ascending hot air. The size of this chamber is critical. This should be as wide at its base as the width of the fireplace and should slope inwards until it reaches the start of the flue (see diagram), 600mm (24") above the smoke shelf.

Flue

The flue carries the smoke, vapours and gases into the chimney and, in a Rumford, the size is critical – the inside area should be one-tenth of the area of the fireplace front opening. For example, 965mm (38") high and 940mm (37") wide, totalling 0.907m² (1,406 sq. ins.) should have a flue area of 0.0907m² (140.6 sq. ins.), and therefore a flue size of no less than 300 x 300mm (12" x 12").

Most manufactured flues – ceramic or metal – come in standard sizes, and therefore it is wise to base the size of your fireplace on the size of the flues available. The inside of the flue should be very smooth, to aid the unhindered passage of the vapours and smoke from the fireplace into the chimney and to the outside. Ceramic or metal ones work well and are readily available.

Chimney stack

The top of the chimney stack should be at least three feet above the roof ridge, nearby trees or surrounding buildings. If your chimney is situated in an area of very high winds, a chimney pot should be placed over the chimney top to discourage crosswinds from billowing down the chimney and interfering with the air circulation.

This Rumford fireplace gives focus to the cosy living space while providing an excellent heat source.

Establishing dimensions

For those of you who hate number-crunching, don't be intimidated by the science involved in creating a Rumford fireplace. Though seemingly overwhelming, the dimensions are fairly simple if worked through methodically. Don't muddle through and try to work them out as you build. If you do, you won't be able to engage with the artistic element of creating a cob fireplace, which can be an opportunity for the sculptural capabilities of cob to truly come into their own.

Constructing the Rumford fireplace

It is easiest to create a Rumford fireplace as you build your walls, but if need you can build the fireplace later, or indeed retrospectively in any type of house.

There are two ways you can go about creating your fireplace, which will produce a different finished design. You can either creating a square fireplace with a lintel, or alternatively, a rounded, arched, beehive-shaped fireplace, which you can make using a home-made wooden former.

1 A fireplace is best built onto a six-inch slab made of lime-crete (NHL 5 - St Astier – 1 part lime : 2 parts sand, stiff mix; see Chapter 5), in order to fireproof the base and to disperse the weight of the chimney and fireplace above it.

2 Once the lime-crete slab has set, spread a half-inch layer of silica sand onto the lime slab, covering the area on which the firebricks will sit.

3 Bed the base of the fire bricks (corresponding to the chosen dimensions of your fireplace) into the silica sand, ensuring that they are all completely level and tightly butted against one another. There should be no gaps between the joints. Use a small rubber mallet to assist with this levelling process.

4 Create the firebrick backing (again, corresponding to the chosen dimensions of your fireplace) stacked three courses high, on their sides. Ensure that there are no vertical joins created between the bricks. The backing should be set with the exact width needed, which may mean that you will have to cut some of the bricks to size. Create the side walls at an angle so that the opening corresponds with the measurement needed to fit into your scheme. Firebricks are used instead of cob for this area, as they are harder, and more resistant than cob to the wear and tear created by the logs.

5 Pack behind the firebrick backing and sides with a normal cob mix, to create a solid base from which to build the rest of the fireplace, and to secure the firebricks in place.

6 Continue building up in normal cob up to lintel height, or around a wooden former. Leave a space the full depth of the firebrick, which will later be built up with the special fire cob mix (see diagram). The front surface of the cob should be left rough to create a good key for the fire cob to attach to.

7 If you are using a lintel, put it in place once you are up to the lintel height, ensuring that the cob is sufficiently dry to take its weight. A reinforced concrete lintel or a steel spanning lintel works well for this. Although it is an ecological compromise, these are the strongest and longest lasting lintels you can use in these circumstances. The lintel should be no more than 130mm (5") deep.

8 Using the special fire cob mix described at the beginning of the chapter, build up the fire-back as seen on the diagram. This should slope gradually forwards, so that when it reaches the smoke shelf height it allows for a four-inch opening between the front of the smoke shelf and the front wall built above the lintel (see point 10).

9 The smoke shelf should be created out of fire cob, and should project out to allow for the four-inch opening mentioned above. It should be roughly 75–100mm (3"–4") deep. It should be situated 300mm (12") above the bottom of the lintel or arch.

10 Build up with normal cob on top of the lintel to create a wall. At 300mm (12")

from the bottom of the lintel, you should be at the height of the smoke shelf on the back wall, and it should correspond with the four inches needed for the opening mentioned in Step 9.

11 The smoke chamber should cove in to meet the flue, and there should be 600mm (24") between the smoke shelf and the beginning of the flue (see diagram).

As you build up the smoke chamber, you will need to apply fire cob (like a plaster) on all sides of the smoke chamber, because when you reach the flue height you will not be able to access this area easily. The inside of the smoke chamber should be as smooth as possible to aid the easy passage of smoke, gases and vapours from the fire into the chimney.

12 The flue will sit snugly on top of the smoke chamber. Make sure that the cob is dry enough to take its weight. You can use standard flues – ceramic or stainless steel – and then pack around it with cob, or simply create the flue hole out of the cob itself.

Refer to previous notes on flues for the specific dimensions. Again, the smoother the inside, the better the passage of smoke and gases from the fire. Build up around the flue in cob until you reach ceiling height. Once you go above ceiling height, the chimney stack should be constructed out of solid masonry or brick. Do not construct the chimney out of cob. Use lead flashing at the seams of the opening to make it watertight.

The dimensions

1 The width of the fireplace opening should be 2-3 times the depth.

2 The height of the fireplace up to the bottom of the lintel should be 2-3 times the depth.

Left: Firebrick backing and bevelled sides. Middle: Using a former to establish a cob arch. Right: Ready for a first firing.

3 The height of the back wall of the fireplace should be the same measurement as the depth. For example: Let's say you decide to construct a fireplace that is 300mm (12") deep. The opening of the fireplace should be 600mm (24") wide, and the fireplace back should be 300mm (12") high.

How to lay a Rumford 'tipi' fire

Because of the extra height and shallower depth of a Rumford compared to most open fireplace designs, the wood is laid in a triangular tipi-style against the back wall in preparation for lighting the fire. This tall fire maximises the space filled within the firebox, which ensures maximum radiant heat output.

1 Lay some screwed-up balls of newspaper onto the hearth. On top of this stack some small, dry kindling pieces vertically so that they stand on their ends, leaning against the back wall. In front of these, stack four logs of well-seasoned dry wood, again vertically on their ends, all pointing up toward the centre of the throat like a tipi.

2 Light the newspaper under the kindling, and also light a sheet of newspaper on top of the wood to heat up the flue and burn up any smoke produced in the initial lighting.

3 As the fire burns the kindling, poke the logs to create a better airflow, which will feed the flame.

4 Keep adding more logs as needed in the same tipi style to keep the fire blazing.

Keeping a good strong, hot flame throughout the duration of the fire will ensure an efficient burn and produce maximum radiant heat. A hot flame will burn off most of the gases produced from the burning wood, which will keep the chimney cleaner.

5 To maintain the fire as it burns throughout the evening or day, just keep adding more logs on end as before. When laid in a tipi style, the burning wood should fall in on itself as it burns, and if laid properly is very stable.

6 Even though the Rumford burns extremely cleanly, it is still a good idea to do a once yearly sweep up the flue. You will notice how little soot comes out.

A Rumford fireplace versus a wood-burning stove

A Rumford fireplace design is so effective that it could be considered comparable to a wood-burning stove with regard to efficiency, effectiveness and how cleanly it burns. Rumford fireplaces and stoves differ mainly in how and what they heat. Rumfords heat through radiant heat, which means the fire warms up the objects it comes into contact with, whereas a stove heats mainly the air around it.

Which one you choose to put in your house will depend on the climate you live in and how often you are at home (and how willing you are to consistently stoke the fire).

Fireplaces are best in less cold, moderate

climates, where temperatures don't fall below freezing for long periods of time. This is because the fireplace needs good ventilation to be effective, which may necessitate the opening of a window to feed the fire in the initial stages and set up the circulation. In very tightly sealed houses, a stove would work better because it is a self-contained system that does not rely on good outside ventilation to operate.

In a very tightly sealed house, and in very cold climates where you wouldn't want to be opening windows, a fireplace would only operate effectively with a separate outside air source that could be closed off when the fire is not in use. For these reasons, fireplaces are traditionally more common in countries with moderate climates such as the UK and in southern Europe. In more northerly European countries such as Germany, Sweden, and in Russia, where houses were traditionally built to seal off the freezing winter conditions, stoves were, and are, more commonplace.

Fireplaces are best used as a top-up to a thermostatically controlled background heat source. This is because, as mentioned before, they only heat what they are in direct contact with, and therefore to heat a house, you would need to have one in every room. On top of this, each fire would need to be constantly burning and therefore tended to regularly. A stove, on the other hand, burns for long hours with minimal tending, and will do its job to create heat in your home even when you are not sitting in front of it.

A fireplace comes into its own as a focal point, and an experience – the smells, the sounds and the feel, an excuse to get cosy and settle down for the evening.

Outdoor Rumfords

Fire and mud can be brought together outside to create wonderful cob courtyard spaces, with a Rumford fireplace incorporated into the walls. The fireplace will effectively heat the bodies around it because of the radiant heat principle, and the thermal mass of the surrounding cob walls will ensure that the heat from the fire is absorbed, stored and then re-radiated back into the courtyard space. If used regularly, this may even help create, along with the sun, a mini- microclimate in the courtyard space, which will help the growth of fruit trees and other plants needing regular warmth.

The construction principles are exactly the same as for an indoor fireplace, but no damper is needed. It is best sited in a fairly sheltered area away from prevailing winds and rain to prevent smoking and damage to the chimney.

The walls of the courtyard and the actual fireplace must be roofed with a suitable material, and the chimney capped with a cowl so that the rain does not cause damage from above. For an outdoor fireplace, you can build your chimney from cob.

If the chimney is in a position to be hit from the side by rain, it should be lime-rendered for extra protection.

Bread Ovens

The creation of an earthen bread oven makes for a great self-contained, simple starter project for anyone who wants to get used to working with cob. It will introduce you to all the vital cob steps, from sourcing the clay, testing for its suitability, collecting materials, making a mix, and gaining experience with the nature of cob as a building material. And at the end of it, you will have produced a functional and hopefully beautiful oven, which will bring you years of enjoyment, and a quality of bread and pizza that far surpasses anything that could be baked in a conventional oven, or bought in even the most prestigious boulangerie.

History
As far back as 4,000 BC, before electricity was even a flicker in human eyes, our ancestors relied on the heat from fire to cook their food. They designed a simple dome-shaped, wood-fired, earth oven with the specific intention of baking bread (pizza came a little later on!). To understand why this design was and still is faultless, and why it will bake a much better loaf than a modern gas or electric oven, one needs to understand a little about the basic science behind the three main different types of heat: how they behave, and how they cook food.

Science
The three types of heat are convection, conduction and radiation. Convection heat is the rising of warm gases or liquids, and the sinking of cooler ones. Modern electric and gas ovens primarily rely on convection heat to cook food. The rising heat is transported unevenly through the air and into the food.

Conduction is the passing of heat from one medium to another, from a warmer object to a colder one. The heat is transported through a solid object and into the food item, such as a sausage cooking in a hot frying pan that is in direct contact with a hot electric or gas ring. This method will only cook the side of the sausage that is in direct contact with the hot pan surface.

Radiant heat is how the heat from the sun travels on to our bodies, the earth, and the plants and trees. Its passage is in even, straight lines, travelling through open space, from a hot object to a cooler one. In an earth oven, the food item is cooked from the heat radiating from the residual heat of the fire stored in the walls (more on this later).

How dome-shaped earth ovens work, and why they bake better bread than conventional electric or gas ovens
The dome-shaped earth oven design bakes such excellent bread (and other things) because it puts into practice all three types of heating: conduction from the contact of the bread dough on the hot fire brick floor, radiant heat being evenly released from the surrounding earthen dome walls, which have stored heat from the fire burning within, and convection from the hot rising air. The oven consists of a dome-shaped structure

fashioned out of cob, which sits on a foundation plinth at a height that is comfortable for the operator to work – shovelling dough in and bread out.

The oven floor is made out of firebricks, onto which a fire is laid from thin pieces of quick burning, dry wood. The simplest oven is created with a single opening at the front, specifically calculated according to the oven size. This opening remains open when the fire is burning, acting as a passage through which air is drawn to feed the flame, and from which any residual smoke can exit.

Naturally, the incoming cold air collects below the exiting warm air and smoke. As the fire burns, it will heat up the firebrick floor it is in contact with, from the hot coals created, and the earthen dome around it – the firebrick and earth both have excellent thermal mass (see Chapter 2), enabling them to absorb and store this heat until the fire is raked out.

When the inside oven space has reached its maximum temperature of around 370°C (700°F), which can take from 1–3 hours depending on the size of the oven, the remains of the fire and hot coals are raked out into a metal container. The oven floor is then wiped with a damp cloth, the oven door closed and the oven left to sit for up to half an hour for the temperature to even out inside and to come down to the necessary baking temperature of 230°C (450°F). The bread dough is then placed directly onto the oven floor, and allowed to cook evenly and deliciously into a crusty loaf. The oven will gradually decrease in temperature over a period of the next six hours as it relinquishes its stored heat.

This makes it possible to cook a range of different foods requiring different temperatures. Utilising these different temperatures is the equivalent to turning the knob on your oven, but a lot less mechanical and a lot more sensory – it involves the

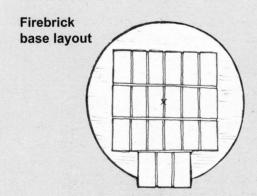

Firebrick base layout

Diagram showing the distribution of firebricks on the oven base.

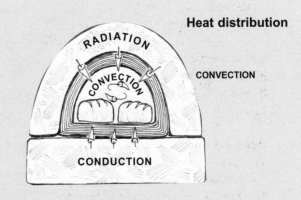

Heat distribution

RADIATION

CONVECTION

CONVECTION

CONDUCTION

Heat distribution diagram of a dome-shaped earth oven, showing the principle of cooking. Drawing inspired by Kiko Denzer.

beginning of a relationship between mud, fire, and your food. At its hottest of around 370°C (700°F), and as soon as you have raked out the fire, you can bake pizza quickly for never more than three minutes. At around 230°C (450°F) it is perfect for bread, and then as it gets cooler, you can bake roasts, pies, potatoes, vegetables and plenty more – it's open to experimentation.

Siting the oven

Before embarking on the construction of the oven, you will need to consider the site on which it is to be built. Though not as critical as when siting a building, it is important to choose a spot that is not too exposed to the prevailing winds, as this will affect the functioning of the oven, the movement of smoke, and how well the fire burns. The oven door should at least be faced away from the prevailing wind. Next to consider is that the oven is a safe distance from flammable materials, as there will often be sparks produced from the fire when it is raked out.

You may choose to use your outdoor oven only in the summer or on fair days, but if you want to be an all-weather baker, it is important that you are not too far from your house, so that you can safely transport your fresh-baked goodies inside on stormy dark nights. The ground on which the oven is built should be level, firm and well drained.

Tools

1 Pegs and string to mark out your oven outline
2 Shovels
3 Wheelbarrow
4 Pitchforks
5 Tarps for making the cob mixes
6 Buckets
7 Tape measure
8 Sculpting tools: spoons, knives, artists' tools
9 Mesh screens for sieving cob to make fire cob plasters
10 Spirit level for levelling firebricks
11 Spray bottle for dampening sand form
12 Pick-axe for creating below ground foundation
13 Rubber mallet for persuading firebrick into position
14 Carpentry tools for making oven door

Materials

1 Stone for the foundation plinth
2 Rubble as a filler for the foundation plinth
3 21 firebricks for oven floor (more or less, if oven is bigger or smaller than the one described)
4 Fine silica sand in which to bed the firebricks
5 Sand for the oven form
6. Normal cob mix with straw for outer insulative layer
7 Fire-cob mix for the first layer
8 Earthen plaster or cow dung to plaster the outside
9 Newspaper to cover the sand form
10 Water source
11 Wood and screws for making the oven door

The oven must have a protective roof covering, as otherwise it will wash away after the first season of rain, as well as making it unpleasant for the baker on rainy days. This can be a simple, free-standing wood structure with a sloping mono-pitch roof (see page 197), or it can become an elaborate outdoor room with benches, a turf roof and kitchen space for bread preparation. Quite often ovens are built under a lean-to at the side of a house.

Regarding the need to get permission from local authorities: in most cases this will only be an issue for those living in urban areas, where smoke control legislation is in force.

Generally speaking, as long as it is a small structure, it will not require planning permission. However, each region within the UK has different stipulations, and if you are at all in doubt you may want to seek advice from your local authority. Building regulations will not apply because your oven is not being built for the purposes of habitation.

The oven design

The actual oven is made up of two layers of differing cob mixes – a no-straw mix for the first thermal layer, a normal cob mix for the second layer, and a finish, breathable plaster, which is best made out of sand and earth, or just cow dung. These are shaped around a damp sand form in the shape of a dome, which is later removed once the cob has dried sufficiently and is able to hold its form.

Oven cross-section

finish plaster

2nd layer

1st layer (thermal cob)

brick floor

sand bed

stone plinth

gravel fill

rubble base

100%   63%

Left: The finished oven

Dimensions

The size of your oven will be determined by what kind of baking you wish to do, how often, and at what volumes. If you plan on baking large batches of bread for example, you will need a large oven to accommodate them. Just one loaf every now and then, and you will want a smaller oven that does not take a long time to heat up and require lots of fuel. A useful exercise when planning the oven dimensions is to mark it all out on the ground and mock up your baking trays or imaginary rounds of bread. See how big the actual oven floor needs to be to fulfil your needs – remember that bread rises and spreads out as it bakes. Once you have determined the oven floor size you would like, everything will develop from there.

A good standard size that we have found meets most people's needs – not too big and not too small – starts with a 690mm (27") oven floor. As the thermal layer should be about 75mm (3"), the second layer from 130–150mm (5"–6"), and the final plaster up to 50mm (2"), you will end up with a total oven diameter of up to 1250mm (49") in total.

- The total diameter of the stone foundation should be slightly larger than the total diameter of the oven.

- The height of the oven void and therefore the damp sand form should be slightly larger than the radius of the oven floor, so the above-sized oven shown in the diagram (page 191), with a radius of 340mm (13 1/2"), should have an actual

height between 430–500mm (17"–20"). If your void is smaller than this, the fire will not burn, as not enough air will be able to enter; and if larger than this, you will get cold pockets in the upper reaches of the void. Both situations will produce less than ideal baking results.

- The opening at the front of the oven onto which the door is fitted should be 63% of the interior dome height. This is a standard and crucial measurement for optimum combustion of the fire within the oven.

The Building Process

1 To create the foundations, dig a hole slightly larger than the total diameter of the oven, going down roughly 450mm (18") into the ground. This is satisfactory for most parts of the UK where winters are not generally too severe. If you live in an area where the ground is frozen for some of the year, you will need to find out where the frost line is and dig down to this level. This will prevent the oven from shifting about and buckling with the frost heave.

Though best avoided, if building on an area that is prone to getting sodden during heavy rain, you will need to construct a drainage trench skirting around the oven (see Chapter 5: *Foundations*). This is not necessary in most cases.

Ensure that the bottom of the hole is solid and compact. If not, tamp with a tamper or large round pole. Fill this hole to 25mm (1") below the ground level with rubble or drain rock about 80mm (3") in diameter, and again tamp until compact. This will prevent any ground moisture from migrating up into the oven base and damaging the cob. On top of the drain rock begin to build the foundation plinth in stone, brick or any other hard material that is resistant to water. It is only necessary to build a stone or brick face – the infill can be rubble. The stacked stone or brick should be at least 300mm (12") wide, and all normal stone/bricklaying principles apply – most importantly that no vertical joints occur between different courses. If your stone or bricklaying skills are a little rusty, use a cob or lime mortar in between the stones to stabilise them. (For information on laying stones with lime mortar, refer to Chapter 5: *Foundations*). Remember that the mud oven is heavy, and the main job of the foundation plinth is to support this weight, so your plinth must be solid.

The foundation plinth can be at any height, as long as it is at least one foot off the ground to protect the cob from moisture. The ideal measurement to go for is the main baker's waist height, so that it is comfortable to work at.

2 Lay your sand base for the firebricks. This base should consist of normal builder's sand about 150mm (6") deep, to help hold the heat in the oven floor and prevent it from being lost into the ground below. This should come level with the top of the foundation ring. On top of this should be spread a 12mm (½") layer of fine silica

sand, which should be extended 12mm (½") above the stone ring around the door area where the firebrick will overhang the foundation slightly. Make a smooth bed, which should be level at all points to prepare for the bricks.

3 The firebricks must be laid from a dead central point so that the floor is set evenly within the oven for even baking. You will need 21 bricks, which will be laid in a square with a tongue at the area where the opening will be placed. With a tape measure, work out and then mark the centre of the oven floor. Lay the first 2 bricks either side of this central point (see diagram on page 191) and then build the others around them.

The bricks should be carefully placed together, making sure that they are very tightly fitted against one another. As you lower each new brick into place, make contact with its neighbour and slowly slide it down into place. Use a rubber mallet to tamp the bricks firmly into the sand – but tamp gently, so as not to disturb the other bricks.

The brick floor must be level, as well as each individual brick in relation to its neighbour – no jutting edges. You may need a few attempts to get it perfect.

Place the three-brick 'tongue' as shown on the diagram (see page 189).

4 To mark the circumference of the sand dome on the firebricks, you need to draw a circle reaching to the farthest edges of the

Above: Creating the foundations for the oven.

Above right: Foundation hole filled with rubble.

Above far right: Finished foundation plinth.

Right: Firebricks for oven floor – note the lip extending out to the edge of the foundation plinth.

Above: The completed inner oven form made with damp sand.

Far left: Building the oven dome of cob around the sand former.

Left: Scoring the surface of the second cob layer to provide a key for the final plaster.

Right: Scraping out the sand form with bare hands.

Far right: Plastering the oven with a cow dung/earth plaster.

bricks. Find your central point in the square, and mock up a large compass using a piece of string attached to a pencil. The piece of string must be the length of the radius of the circle (half the diameter).

Start with a pile of damp sand in the centre of the circle, building up and out so that the edges of the sand form stop at the edge of your pencil marking, so that only the corners of the bricks are visible.

Build the walls almost vertically at first to create enough space within the oven, especially at the edges, for the bread dough to rise unhampered. At about 115mm (4 1/2") begin to curve the sand in to create a dome, and build up to height 495mm (19 1/2") for the oven size described earlier. Use a stick marked with this height, and a level as a guide.

Use your hands or a piece of flat wood to make your dome compact and smooth. If building with cob takes you back to the worry-free days of making mud pies, building a sand dome will take you back to castle building on the beach. It is all good fun. Start off with the sand slightly on the dry side, as it is easier to spray it with more water, than to wait for sloppy sand that won't hold its form to dry.

While the sand dome is still damp, cover it with damp newspaper so that when you are digging out the sand you will know when you have hit the beginning of the cob walls. It will be dark in there!

5 Make up the two separate batches of cob: one without straw for the thermal layer, and one with straw for the second layer. Refer to Chapter 4: *How to make a cob mix.*

Mark out on the foundation base the 75mm (3") needed for the thermal layer, and then the 150mm (6") for the second layer, to guide you.

With handfuls of fairly stiff, strawless cob, compress it in place starting at the base of the sand form. The cob should be pressed into itself and not into the sand form; otherwise it will damage the form, causing it to lose its shape. Maintain a good edge as you build up. The dome must be covered in one go to avoid joins and thus weak spots in the oven structure. When completely covered, do not slap the cob with your hands, as it will weaken it, but rather use a flat board to compact it, gently rocking it over the cob until it is firm and smooth. With a fork or plastering scratch tool, score the surface to create a key for the next layer.

The next layer of cob (with straw) can be applied immediately. If you decide to wait some time before doing this, and the first layer of cob has dried, it should be dampened before applying the second layer, to ensure good bonding. Apply the second layer in exactly the same way as the first, leaving a scored surface to key in the plaster layer.

6 Cut out the door and extract the sand. Measure out the door height and width

(height = 63% of the total dome height, i.e. 500mm x 0.63 = 315mm (19 $^1/_2$" x 0.63 = 12 $^1/_4$"). The width should be wide enough to be able to fit loaf tins and pizzas, but not so wide that too much heat is lost. We recommend a width of about 500 mm (19 $^1/_2$").

Wait till the cob feels firm to the touch before carving out the door hole. Scratch the outline of the door measurement with a metal tool – a dull knife or trowel – and then slice through the cob layers to create the opening. Take a deep breath and begin pulling out the sand. Use a flashlight to see the farthest reaches, and to see when you have hit the newspaper, which should be peeled off the cob and removed. Have faith: the cob dome will hold its shape if the right steps have been followed.

7 As with all earthen structures, the oven must be covered with a breathable plaster. This is especially pertinent with the oven, as it will need to release the steam created from the heat of the fire. A lime plaster can be used, but we prefer to use something less processed, so as to retain the muddy tones of the cob, and to remind us that it is built of the earth around us – dug up from our back garden or an adjoining field.

An earthen plaster or straight cow dung plaster mixed with sand seem the most appropriate options, and the most fun! For details on creating and applying earthen and cow dung plasters see Chapter 9: *Lime & other natural finishes*.

The oven needs to dry for at least a few days before laying the first fire, to ensure that the intense heat does not create cracking in the cob.

A nice way to christen your oven, and to be able to appreciate the magical inner dome that you have created out of mud, is to light a candle inside the structure. Watch the flame dance on the mud.

The door
The door should be made out of wood to fit snugly into the opening. It is necessary to have a handle on the door to make it easy to put in place and to remove. It does not need to be hinged in place.

The oven and smoking
Be aware that the oven may smoke a little – don't worry, as it is all part of the process. Because the earth oven as described does not have a chimney, the fire may not burn as efficiently as it should. This may only happen as you are getting the fire going and the oven is cold. As long as dry wood is used and the fire burns well, smoke should be kept to a minimum. This is why it is best not to enclose an oven in a building, and why the oven door should be placed away from the prevailing wind.

If you really want to go all-out, or build the oven in a well-ventilated indoor space, it can be designed with a chimney to deal with the smoke. However, we find that the simple design described in this chapter works superbly.

How to lay the fire & use your earth oven

This takes practice and experience, but there are a few guidelines that can get you on your way:

1 Use dry, thin, soft or hardwood. Scraps of wood can be used to good effect as long as they are not impregnated with poisonous chemicals, or covered in lead paints.

2 The wood will burn best if it is stacked loosely to allow lots of air to fuel the flames.

3 Build the fire at the front of the oven first, and gradually push it back over the course of the firing.

4 Allow the fire to burn for at least 30 minutes with the door open and then feed it with some more wood – to encourage a good burn.

5 Allow to burn from 1–3 hours until the oven reaches its maximum temperature of around 370°C (700°F). Leave a baker's temperature gauge in the oven which will be able to record higher temperatures when reached.

6 Just before the oven reaches its peak temperature, spread the coals and embers across the whole floor of the oven to heat the firebricks evenly, as most of the heat will have been concentrated under the fire in the centre of the oven.

7 When ready, rake out the remains of the fire into a metal bucket (anything plastic will melt). Sweep the firebrick floor or wipe with a damp cloth so that you don't get soot or grit in your bread.

8 Shut the door, and allow the heat to even out and really soak into the structure – about a half an hour is good. If baking pizza, however, cook immediately after the fire has been raked out as they cook best in very hot temperatures.

9 Monitor the temperature gauge and set in the dough quickly as soon as it lowers to 230°C (450°F). Shut the door immediately to prevent heat loss. The bread cooking time will vary due to your recipe but should be ready from between 20-60 minutes. You will get to know through trial and error – that's the fun!

The baking process using an earthen oven is an almost full-day affair, which can be turned into a ritual, and an event in itself. It truly produces 'slow food', which tastes wonderful, and is much needed in today's hectic and fast world.

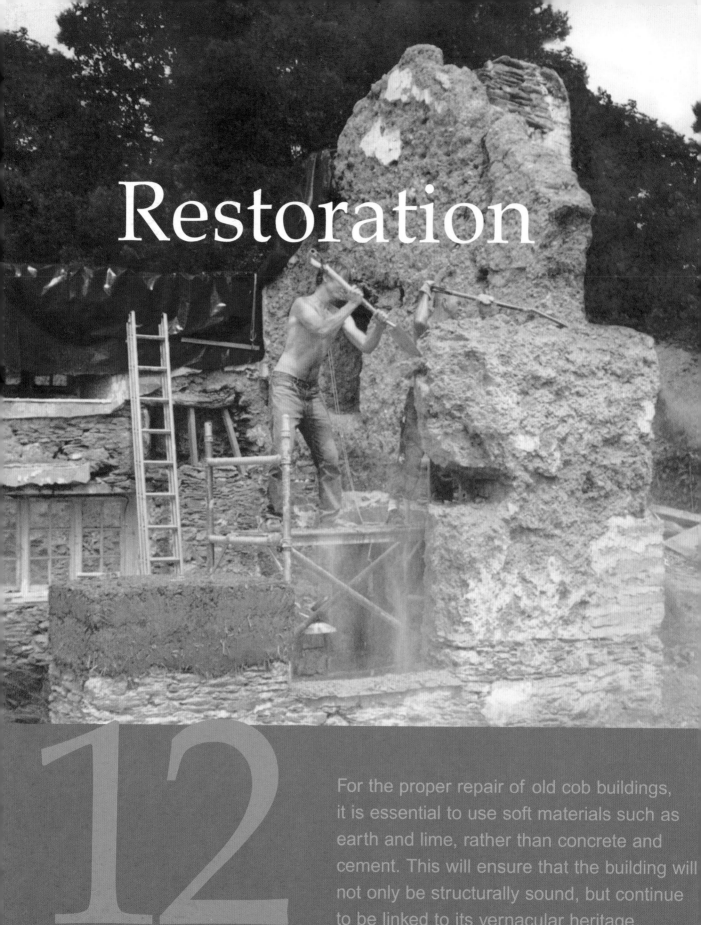

Restoration

12

For the proper repair of old cob buildings, it is essential to use soft materials such as earth and lime, rather than concrete and cement. This will ensure that the building will not only be structurally sound, but continue to be linked to its vernacular heritage.

Chapter 12

In recent years, there have been considerable changes in the approach to restoring old properties. Thankfully we are moving away from the hard language of the 1960s, 70s and 80s, of cement and concrete, hard straight lines, mechanisation, mass-manufactured materials, and a need to bring order and uniformity to our rambling old buildings. In the twenty-first century we are seeing a new paradigm emerge, that speaks a softer language: of lime and earth, the reinstatement of soft, undulating forms, hand-made materials and the confidence to allow uniqueness and imperfection to once again shine through. As important as our philosophical approach to cob buildings, and the new aesthetics of how a sensitively restored cob building looks, is the practical issue of using like-for-like materials to produce a more structurally sound building. Consider the bond between a hard, manufactured concrete block, or brittle cement render, with the soft earth of an old cob wall. They don't make good partners. It would be equivalent to covering your smooth skin with mud – when it dries, it would crack and fall off whenever you moved.

At last, it seems that many people are beginning to rediscover these ideas, and all over the UK you can hear the collective sighs from old cob buildings as they are liberated from their cement armour, and literally begin to breathe again. These breathing cob buildings, relieved of their hard coats, reveal their history, and

Left: Cob wall ready to be re-rendered. Middle: Cob Quaker Meeting House. Right: A fully restored inglenook.

hence the stories of the people who made them and the communities in which they lived. Old buildings contain valuable information about our history, and this is particularly important with regard to cob buildings, because they were the work of the rural builders, farmers and 'ordinary' folk who worked intuitively, and not of architects or highly skilled craftsmen. For this reason there are no official written records about the ancient techniques of cob building, and it is therefore important to restore these old cob buildings as true to their original form as possible. To achieve this, it is vital that the appropriate methods and materials are used to repair and maintain them. This way the links to the past will be maintained and kept alive well into the future.

There is no doubt that cob buildings need a little more tender loving care than a house constructed out of concrete blocks. It is absolutely vital to keep up regular maintenance measures to prevent excess amounts of water from penetrating the walls and causing damage. These involve simple tasks such as the regular clearing of leaves from the gutters, mending leaks in the roof, and carrying out springtime lime washing. If you stick to these regular routines, your old cob building will last for as long as it's needed.

The following information in this chapter runs through a comprehensive list of potential problem areas that could arise in an old cob building, and explains how to remedy them. Before commencing on any repair work, make sure to ascertain the original source of the problem and repair this first. For example, a repaired structural crack in a cob wall that has appeared as the result of the penetration of water from a hole in the roof will simply reappear if the hole is not first mended.

Most of the repairs outlined below can be carried out by someone who has little

tools

Tools for restoring cob

1. Skutch
2. Lump hammer
3. Churn brush
4. Plastering/rendering tools
5. Paint brushes
6. Buckets
7. Old saws
8. Mini mattock
9. Cat's claw/crow bar
10. Masonry chisel
11. Garden sprayer
12. Pitchforks
13. Shovels
14. Cob beaters
15. Carpenter's wooden mallet
16. Levels in various sizes

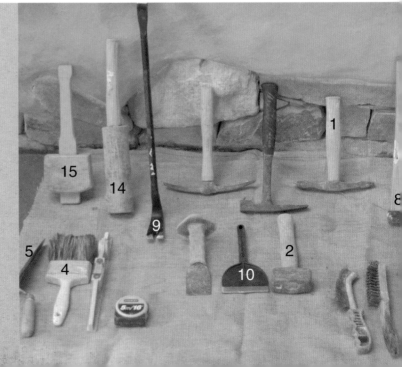

experience of building work. The only area that we would express caution with is in the repair of large structural cracks, which can be defined as a crack that is larger than 25mm (1"), especially if they appear to be growing, and any damaged areas that appear to be jeopardising the structural integrity of the building.

Consult someone who has experience of working specifically with cob buildings. If there is no one in your area, contact an organisation which specialises in the repair and conservation of old buildings such as SPAB (Society for the Protection of Ancient Buildings), DEBA (Devon Earth Building Association), EARTHA (East Anglia Earth Buildings Group) or English Heritage (see *Resources* for contacts). Be wary of conventional builders, who may tempt you to carry out repair work using unsuitable modern materials such as concrete blocks and cement, and too many plumb lines!

If your building is listed, or located in a conservation area, find out from your local conservation officer whether there are any special requirements when you carry out repair work on the building. If you are choosing to restore sensitively and with the appropriate materials, they will be extremely supportive of any repair work that you wish to do.

Using pre-dried blocks for restoration

Pre-dried cob blocks are an essential item to have at hand when repairing an old cob building. As described in the following points, they are used in a variety of areas, such as repairing structural cracks and building out window and door reveals. The main advantage of using pre-dried cob blocks is that they eliminate the

Pre-dried cob blocks (left), which can easily be made using a wooden former (right), are an essential element in the work of restoring old cob buildings.

problem of shrinkage that occurs when new mass cob dries against old cob. This shrinkage will create a fault line, which may cause a vulnerable area in the future and hence fail to remedy the problem in the long term.

Cob blocks are also useful when large areas of a cob wall need to be built up in a short period of time, because they avoid the drying period needed between lifts of mass cob. For this reason too, they are indispensable when building work must be carried out in the winter, when cold and damp weather will hinder the drying of mass cob and make progress slow. In areas that are difficult to access, such as where the top of the wall meets the roof, cob blocks can provide a solution.

Cob blocks can either be made at home with a simple wooden mould, or bought from a variety of suppliers. The latter option is fairly expensive, and of course incurs transportation, but cuts down on the labour and time involved in making them yourself.

They can be made and bought in a variety of sizes – the two standard sizes being the same as a modern concrete block and brick. Cob tiles (which are much thinner than a cob block or brick) are also good for stitching cracks, as they offer a high degree of flexibility and are easy to handle. If you make up your own, you have the advantage of being able to tailor-make them to the sizes you want.

How to make a cob block

Make up a simple wooden former of appropriate depth: either a single mould or a ladder system containing up to four sections (see picture on page 201). Protruding wooden handles at each end of the ladder allow two people to shake the blocks out of the mould.

1 Prepare a suitable site where the cob blocks can be laid out to dry. It needs to be well ventilated, but protected from the rain and direct sun – a large barn or polytunnel is ideal. A series of wooden planks (scaffolding planks work well) need to be laid out on level ground onto which the blocks will be placed; or a clean, level floor will suffice.

2 Prepare a suitable cob mix (for details refer to Chapter 4: *How to make a cob mix.* The cob from an old wall can be re-used to make the blocks, with the addition of fresh straw. The cob should be made on the stiff side, to minimise cracking and so that the blocks can hold their form when they are removed from the mould.

3 Wet the mould. Place the mould on a wooden plank or on clean, level ground. Wet the mould thoroughly so that the blocks will slide out easily.

4 Pack the mould with the stiff cob mix, ensuring that the corners are tightly packed with material. Use the weight of the bottom of your foot to really compress the cob into the mould. Build up the cob in layers within the mould until it reaches the top,

Making cob blocks

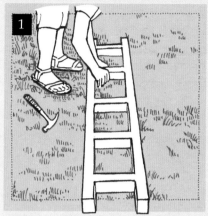

1

Make a mould suitable for your purposes.

2

Prepare a stiff mix.

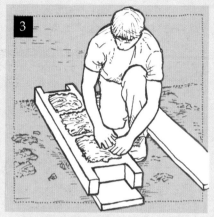

3

Pack wetted form tightly with cob mix.

4

Place filled former on prepared surface.

5

Shake mould to release blocks.

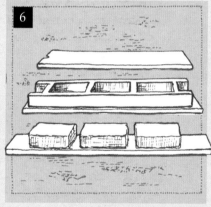

6

Leave blocks to dry.

Prepare a suitable site where the cob blocks can be laid out to dry. Old scaffolding planks work well for drying blocks. Make your former.

Prepare a suitable cob mix. The cob should be on the stiff side, to minimise cracking and so that the blocks can maintain form when they are removed from the mould.

Wet the mould. Place the mould on a wooden plank or on level ground. Wet the mould so that the blocks will slide out easily.

Pack the mould with the stiff cob mix, ensuring that the corners are tightly packed.

Shake the mould, with one person at either end of the mould.

Allow the blocks to dry on their flat sides, until they can be safely moved.

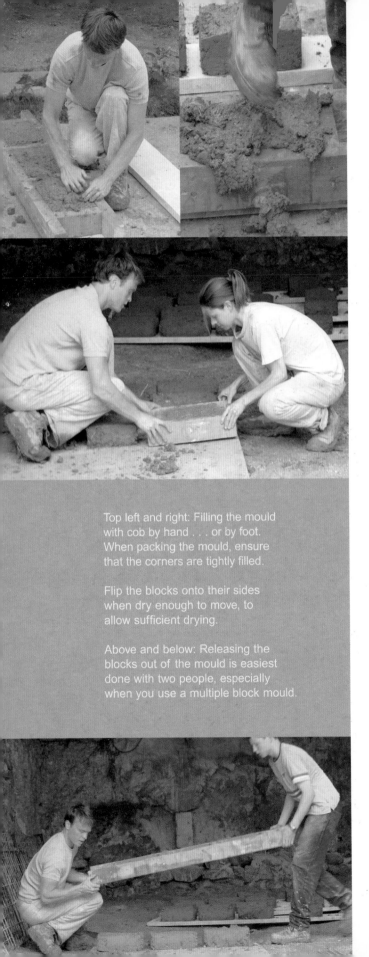

Top left and right: Filling the mould with cob by hand . . . or by foot. When packing the mould, ensure that the corners are tightly filled.

Flip the blocks onto their sides when dry enough to move, to allow sufficient drying.

Above and below: Releasing the blocks out of the mould is easiest done with two people, especially when you use a multiple block mould.

so that the end product is a really well-compacted, dense block.

5 Shake the mould. With one person at either end of the mould, grasp the wooden handles and shake vigorously until the blocks slide out onto the board. Keep the mould low to the ground as you shake the blocks out, as otherwise they will be damaged as they fall onto the board. If the blocks do not slide out of the mould easily, your mould was not wet enough before you packed it with cob. Wet the mould thoroughly between each batch of blocks.

6 Allow the blocks to dry on their flat sides until they can be safely moved, at which point they should be tipped onto their sides until thoroughly dry. This can be anywhere from three weeks to three months, depending on the weather and the amount of air circulation there is available.

Cob blocks can be laid into either a lime or cob mortar. We prefer to use a cob mortar because of its compatibility with the cob block and hence its ability to bond extremely well. It also has the benefit of being more cost-effective than the lime. To make a cob mortar, simply mix water with a normal cob mix to produce a mushy consistency. If you choose to use a lime mortar, a standard sand/lime mix should be used (see Chapter 5: *Foundations*). Always wet the block before laying, either by dunking it quickly into a bucket of water, or brushing it with a wet paintbrush. This will ensure a strong suction between it and the mortar.

Typical problems that you may encounter in old cob buildings

Large structural crack at the corner of the wall (larger than 25mm/1")

Cracks can often appear at the corners of old cob buildings for a number of reasons. The walls can be forced out by the roof timbers owing to the absence of a sufficient tie across the trusses, which can in turn be caused by rotting or fracturing in the timber frame of the roof structure. Additionally, cracks can occur when there is shifting in the stone foundation plinth caused by differential ground settlement or a poorly built plinth; for example, where no tie stones were used. If you observe a fault in the stonework, this indicates that the latter could be the cause. Otherwise, cracks in the cob walls are most likely to be due to failure in the roof structure. Remember, though, that often if there are serious structural problems in a cob building, it is normally not due to one reason but many different factors, so all possible causes should be investigated.

If these cracks are large and appear to be getting larger, they may cause eventual structural failure. It is for this reason that in this area we recommend that you seek the advice of a professional cob builder, or a structural engineer who has experience of working with cob. To monitor whether a crack is growing, it is possible to use a 'tell-tale' crack-monitoring device that can be purchased at any builders'-merchants. If

using this method, it will need to be monitored for one year. In the meantime, the crack should be temporarily filled with stones and lime or cob to prevent water penetration through the crack. The standard repair treatment for corner cracks is a corner cob block stitch (see diagram on page 207), which involves tying the two separating walls together using pre-dried cob blocks. Special reinforcing rods such as a Helibar (explained on page 208) can be used between the cob block joints to further strengthen the corner, but are not always necessary.

1 Starting at the lowest point of the crack, chase away the original cob in order to make a flat bed to lay down the first full-length cob block. Only cut enough of the original cob to make room for the first course of blocks so that they fit snugly into the excavated hole. Do a trial fit before committing the block, allowing a little extra space for mortar.

2 Thoroughly wet the original cob using an old paintbrush dipped in water and flicked onto the wall, or a garden sprayer if the cob is especially friable.

3 Lay down a bed of cob or lime mortar onto the chased out area, ensuring that the mortar will come into contact with all parts of the block touching the old cob.

> It might seem an easy proposition to knock down an old cottage, particularly one built out of cob. But no! . . . A demolition man we know spoke despairingly of his experiences pulling down cob structures with three-foot thick walls: he threw a steel hawser round the house, attaching it to his tractor and hauling away, confidently expecting the poor old ruin of a house to subside into the dust; only to find that the hawser cut clean through the walls, which dropped an inch and settled comfortably back into position.

Robert Edmunds, *Your Country Cottage: A Guide to Purchase and Restoration.*

4 Dunk the cob block to be laid in a bucket of water to allow for good suction between the mortar and the block.

5 Immediately fit the block into the chased out area, ensuring that the block is tightly butted against the original cob. Ensure also that the exterior face of the block is flush with the face of the original cob wall.

6 Complete first course of blocks. Fit the remaining blocks necessary for the first course to create a solid base onto which the blocks inserted above will rest.

7 Fit blocks to full height of crack. Continue moving up the crack inserting the blocks on top of each other, each time into a bed of freshly laid mortar until the top of the crack is reached. It is vital to ensure that no vertical joins are created (bond the blocks like bricks) and that the blocks are keyed into the original cob wall by stepping each successive layer. Ensure that all blocks follow the vertical face of the wall even if the cob wall is not plumb.

Some helpful pointers
- Use stones and bits of broken cob block bedded into mortar in any voids not filled by the cob blocks.
- Cob blocks can be cut to size to fill smaller spaces using a skutch or lump hammer and masonry chisel.
- You will create a stronger corner if the blocks are placed to fit the full width of the wall, i.e. rebuilding both exterior and interior faces.

Vertical crack in a cob wall
This involves stitching the vertical crack with a succession of horizontally laid cob blocks at 500mm (19$1/2$") intervals.

1 Starting at the bottom of the crack, chase out a hole in the original cob to fit a cob block. Roughly fit a block into the excavated area as a trial run to ensure the block fits snugly, allowing for extra room for mortar.

2 Thoroughly wet the original cob using an old wetted paintbrush, or fine mist sprayer if cob is friable. Lay down a bed of cob or lime mortar into the chased-out area, ensuring the mortar will cover any part of the block that comes into contact with the original cob wall.

3 Dunk cob block to be laid in a bucket of water to allow for good suction between the mortar bed and block. Immediately fit the wetted block laid horizontally into the chased out area. Make sure the cob block is tightly butted up against the mortar and original cob; if not, use stones and mortar to pack out. Ensure also that the face of the block is flush with the face of the wall.

4 Pack out rest of crack with blocks or stones. Depending on the size of the crack you can either fill the space between the horizontally laid blocks with vertically laid cob blocks, or stones and small chunks of cob blocks stuffed into mortar. The former may be necessary for wide cracks.

5 Fit blocks up to full height of crack at intervals. Continue moving up the crack spacing horizontal cob blocks at 500mm (19 1/2") intervals, filling in between until the top of the crack is reached.

If the crack spans the full thickness of the wall, a repair will be needed to both the

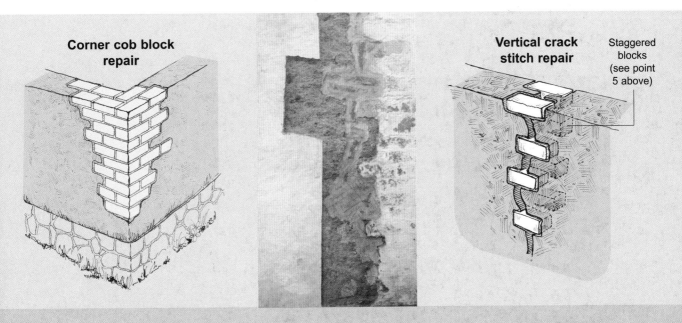

Corner cob block repair

Vertical crack stitch repair

Staggered blocks (see point 5 above)

Above: A completed corner repair showing correct overlap of block courses.

Middle and right: Examples of vertical crack stitch repairs using cob blocks. The spaces in between the cob blocks can be stuffed with stones embedded in lime or earth mortar.

internal and external faces of the wall. The same method can be employed on both faces, except that the horizontally laid cob blocks must be spaced at different heights on either sides of the wall so as to avoid chasing out too much of the wall in one area.

The use of Helibar for corner and stitch repairs

Helical ties (see photos on page 213) can be used to further strengthen a cob block repair. Helibar (made by a company called Helifix) is a twisted stainless steel reinforcing rod which can be used between the joints of cob blocks to tie them into an existing cob wall or any other wall fabric, such as brick, wood or concrete. The corkscrew twist in the metal is ideally suited to cob because it creates a good bite as it enters the cob and thus provides a really good key. They are also good for cob repairs because they allow for some flexibility, which is vital for any material that is introduced into a naturally shifting cob wall. They are also resistant to corrosion from the naturally

occurring moisture in the cob, and have a high tensile strength. You can either use a 'dry fix power driver' piece that is attached to an SDS hammer function drill, or a much easier route is to pound it in with a lump hammer. 400mm (16") lengths are good to use for cob block repairs. If you have bought larger lengths, use bolt croppers to cut them to the desired size.

Lintel replacement

Traditionally, a good hardwood was used for the exterior lintel, such as oak or sweet chestnut, and a softwood for the interior, such as pine. If any of the existing lintels are rotten, it is strongly advised to carefully replace them, using acro-props (adjustable, temporary load-bearing supports) for extra support if the cob appears unstable.

If the cob seems solid and in good condition, lintels can normally be replaced using no extra support. This is an area where you may want to seek professional advice if you are at all uncertain about the stability of the cob.

Careful removal of a rotten lintel within a cob wall. Ensure that the load above the existing lintel is secure before replacement.

New lintel safely in place. Note cob brick fill above the centre where cob had crumbled when the old lintel was removed. See the finished lime-plastered fireplace on page 199.

1 Chase out the lintel on one side of the wall, and immediately replace with a new lintel. Pack with flat stones or slate and lime or cob mortar if necessary.

2 On the other side of the wall, remove the outside and middle lintels and immediately replace, packing out as before. If extra support is required, you can acro-prop the middle lintel immediately after insertion, and before inserting the outside lintel.

Mass cob replacement of a section of cob wall

If an area of damaged cob wall is sufficiently large, or a wall is severely leaning outwards, it is best dealt with by being carefully dismantled and replaced with new mass cob. There are two methods that can be used here to ensure that there is a good bond between the new cob and the old cob, and to avoid the possibility of a separation occurring as the new material dries out and shrinks back. Either the new cob can be tied into the old cob using pre-dried cob blocks (described below), or the remaining cob can be pared down to create a roughly 45 degree angle, into which the new cob will sit.

1 Carefully chase out the area of damaged cob wall.

2 Create steps into either side of the remaining cob, into which you will slot pre-dried cob blocks.

3 Thoroughly dampen the old cob and begin building up new mass cob in

suitable lifts until the top is reached, slotting dampened cob blocks into a bed of cob mortar as you build up the new section.

The great thing about cob is that it can be eternally recycled. Any old cob that has been salvaged from a collapsed wall, or a section in need of repair, can simply be remixed to build the new section. It is necessary to add some fresh clay and some new straw to produce a good strong mix.

Erosion where the cob meets the stone foundation plinth

Erosion can occur at the interface between the bottom of a cob wall and the top of the stone foundation plinth. This can occur if the cob is unprotected and is receiving driving horizontal rain or wind. It can also occur when large farm animals such as cows rub up against the wall to scratch and lick it like a salt lick. If the erosion is under 250mm (10"), it most likely poses no structural problems and can easily be repaired using cob blocks and mortar with Helibar, if extra anchoring is necessary. Anything over 250mm (10") should definitely be seen by someone who has experience with old cob buildings, as rebuilding the wall may be necessary.

Rebuild a damaged stone plinth

If the stone foundation plinth has been damaged and needs repair, rebuild stone using a lime mortar up to original height, creating a solid base from which to place new cob blocks onto. Chase out the necessary cob to make a flat platform for the cob block to rest on.

Top left: Erosion can often occur at the interface between the stone foundation and the bottom of cob wall.
Top right: Chase out the cob to make a level platform.

Middle left: Apply a bed of cob mortar.
Middle right: Fit dampened cob blocks. Ensure that all blocks remain level as you build up.

Below: Erosion under 100mm (4") deep, such as in this window reveal, can be built out in lime plaster.

1 Cut and shape the cob blocks to fit the profile of the eroded wall using a skutch or a lump hammer and masonry chisel (make sure you wet down the block before shaping).

2 Thoroughly dampen the old cob with water before building up the eroded wall face with shaped cob blocks. Bed the blocks into an earth or lime mortar, and make sure the mortar is placed where the block butts up against the old cob wall face. Ensure that the blocks remain level with the stone plinth as you build up.

3 To provide extra anchorage between the cob blocks and the existing cob wall, Helibar (see pages 208 and 213) can be fitted between the cob block joints and driven into the old wall with a lump hammer.

Note: It is always necessary to cure the original cause of the erosion to prevent it from happening again. This may involve rendering the wall with a lime render, and/or preventing animals coming close to the wall. This will be aesthetically necessary anyway to visually tie the cob blocks in with the rest of the wall.

Eroded door and window reveals

A reveal refers to the side of the opening in the wall for a window or door.

Window and door reveals are vulnerable spots which are prone to erosion, especially in neglected barns and dwellings. There are two repair strategy options, depending on the severity of the erosion.

Erosion under 100mm (4")

Build out to the desired shape in consecutive layers of scratch coat lime render/plaster.

1 Take the original cob back to a place where it feels stable.

2 Dampen the surrounding cob thoroughly to ensure a good key.

3 Apply consecutive layers of lime render/plaster, allowing each coat to dry in between, until you reach the desired reveal form. Do not apply each layer in excess of 25mm (1").

Do not try to achieve a perfect line – one of the beautiful features of old cob buildings is their rounded, undulating reveals.

If the erosion is severe

If the reveal is no longer providing a good structural foundation for the lintel, the reveal will need to be built up in cob blocks to restore its structural integrity.

1 Chase away the existing cob to a solid point and create a series of steps in the cob large enough to receive a snugly fitted cob block.

2 If the cob is chased away to a point where the existing lintel is undermined, acro-props will be needed to support the lintel until the repair is complete.

3 Dampen the existing cob and place a bed of cob or lime mortar on the step.

4 Dunk the cob block in water.

5 Lay the cob block on the bed of mortar, ensuring it has a flush face to both the existing cob wall and the profile of the reveal.

6 Continue laying the cob blocks into mortar until the top of the reveal is reached.

7 Lay each consecutive layer of cob blocks in opposite directions so as to avoid a fault line in the joints.

Rats and mice

Rats and mice are attracted to the unthreshed grain from the straw in a cob wall. In poorly maintained buildings they can cause havoc by creating holes and tunnels in the walls. If there are only a few tunnels it is possible to fill them in by bedding cob blocks or stones into a lime or cob mortar. If there are extensive tunnels, a honeycomb effect will be created and the structural integrity of the wall could be undermined. It may therefore need to be taken down and rebuilt either in mass cob or cob blocks. If it involves bringing down the corner of a wall, it is best built back up in cob blocks.

1 Before carrying out any repairs, ensure that the rat/mouse problem is solved, to avoid a potential recurrence of the same problem.

2 Wear gloves to carry out any work where rats may have been present.

3 Remove any loose material from the tunnel – you may find remnants of rats' nests – as far back into the wall as you can reach.

4 Dampen the hole thoroughly and either plug the hole with cob blocks and stones bedded into a lime or cob mortar, or alternatively use mass cob to plug up the hole, building it out in at least two layers depending on the size of the hole.

Our experience with the second method has been less satisfactory than the first, due to the shrinking and lifting of the new cob from the old, and a consequent loosening of the new material.

Masonry bees

Masonry bees love to burrow into cob walls, especially old cob which is un-rendered and has become friable and easy for them to penetrate. This is primarily an aesthetic issue, and rarely will cause any structural damage.

To remedy the problem, the wall concerned can simply be rendered in lime or earth, but make sure that no bees get trapped in the wall.

Cows rubbing up against cob

See the above point dealing with eroded cob walls at the interface of the stone plinth.

Plant life

Any plant life that is growing from the walls should be removed to avoid damage. Neglected cob buildings are often found with ivy growing up them. If it is extensive and has been there for some time, the roots have the potential to burrow deep into the wall and eventually cause some weakness. This is especially true if the wall is not rendered or lime washed.

To remove the ivy, carefully cut back the leaves and remove the surface roots. It is very stubborn stuff, and it is therefore essential to kill off the roots to fully eradicate it before rendering or lime washing.

Having said this, we have seen many cob walls, which have clearly been covered in ivy for many years and are very happy – the ivy even seeming to afford the wall some protection. If you have a boundary or garden wall, or ancillary building that is covered with ivy, it should be sufficient to just keep it closely trimmed.

Depressions in the surface of a cob wall

When repairing an old cob wall and preparing it for plastering/rendering, it is common to find surface depressions where the cob has been eroded over the years. To fill these holes there are two options:

1 For shallow depressions, a lime render mix can simply be thrown into the pre-wetted hole and built out in layers. Allow each layer to set in between.

2 For deeper depressions, fill with cob blocks as in the point above which deals with rat/mouse holes.

incorrect

Laying brick courses

correct

Above: Always ensure that successive courses of blocks are staggered to tie the wall structure together.

Top left and middle: Trimming cob block to size.

Top right: Wetting cob block thoroughly before placement.

Above left and middle: Also wet the surface where the cob block will be placed, and firmly press into position.

Right: Using Helifix to tie cob block repair into the existing cob wall.

Bottom right: The Helifix tie is sandwiched between the cob blocks.

Above: The cob blocks are bedded into a mortar of cob or lime.

Dampness in old cob buildings – removing impermeable wall coverings and re-covering with breathable finishes.

In Chapter 9: *Lime & other natural finishes*, we have talked at length about lime and cob, and why cement and other impervious renders, plasters and paints are detrimental to cob walls. It is because they fail to allow them to breathe. If you experience damp problems in a cob house, we urge you to read Chapter 9.

Most damp problems will be remedied by removing all impervious finishes. This includes the external pointing mortar in between the stones of the foundation plinth. There are other areas of the building that may be creating moisture problems in the walls, and these should be investigated and remedied before carrying out any plastering or rendering work.

Possible causes of moisture ingress may include:
- Failed roof coverings, such as slipped tiles and slates or neglected thatch.
- Defective lead flashings, leaking gutters and drains.
- Earth built up against the cob wall.
- Poorly designed windows and doors such as defective window sills.
- Defective drainage – always ensure that all water is drained away from the building. Gravel placed at the base of the wall will prevent the splashback from rainwater.

Removing cement and replacing with lime

Removing cement render from a cob wall can be a fairly taxing task, but the effort made will be richly rewarded. You will not only end up with a healthy and sound building, but aesthetically you will be restoring the characteristic soft forms of cob that are often disguised by hard cement render.

Some areas of cement may have separated from the cob wall, and these will sound hollow when tapped. This is a good place to start the removal process, and they will be the easiest areas to remove, often taking only a good tug to pull away large chunks of render. Take care and protect yourself from falling render, as more can come away than you bargained for. Areas that are more strongly bonded to the cob will take a lump hammer and masonry chisel, a skutch or even a crowbar to prise the render off. Start where there are cracks evident in the cement, moving along the wall until it is all removed. Remove the cement very cautiously, as the cob underneath is likely to be friable and may pull away with the cement. A thick cement coating can hide a cob wall which has been in a very bad condition for years. Any particularly stubborn areas of cement may pull too much of the cob away when removed, and are best left in place and incorporated into the overall lime coating. You may expose cracks in the cob that have formed underneath the cement. These will need to be repaired using cob blocks, using the techniques explained at the beginning of this chapter.

Once the cement has been removed, you may also expose large areas of cob that have been damp for some time. These areas need to dry out before applying a new, breathable finish – you will notice the change in colour and texture when it is ready. Any mould or mildew that is evident on the cob will be remedied by the exposure to the air and a covering of lime, which deters fungal growth due to its high alkalinity. When the wall is sufficiently dry, apply a new breathable finish (see Chapter 9 on good application practice of natural finishes). Old cob walls will generally need three coats: an initial 'dub' coat, which involves filling in any voids and/or surface depressions in the cob; a scratch coat; and a smooth finish coat for the 'dub' coat. Simply throw balls of lime into the voids. It is not necessary to trowel them in as the rougher it is, the better the surface key is for the next coat. Any large voids can be dealt with as described above.

When applying the scratch coat on extremely crumbly old walls, a harling or throwing method should be employed, or a tyrolean machine or render gun used. Application with a trowel will apply too much pressure to the wall and may end up pulling off vulnerable areas of cob. For information on using a render gun or tyrolean machine, see Chapter 9. The harling method involves using a special harling tool (or a shortened coal shovel will do), which is used to flick a lime sand slurry (1 : 3 lime sand mix – more water than a normal mix), using a firm hand movement to get a good key. The same mix is used as for a normal lime plaster or render, only it is made slightly wetter. This can either be hard trowelled over, or left rough to give a good key for the next coat. A final covering of lime wash, especially with pigments, will greatly enhance the beauty and character of the building.

Removing cement pointing mortar from external stone joints

To remove the cement pointing mortar in between the stone joints, simply rake out with a mini-pick, and replace with a lime mortar. Please refer to Chapter 5.

Consolidation of an existing plaster or render

There may be situations where the existing lime coating is in a good enough condition to keep, but has areas that are cracked and flaking. In this instance, the failing areas of lime can be carefully removed and the exposed area patched with new lime coatings. To blend the old into the new, it can be feathered in, using circular strokes with a damp sponge when the plaster/render is green hard. You may need to build it up in layers.

Planning permission & building regulations

13

Contrary to popular belief, a well designed new cob building can comply with the rigorous standards of the building regulations.

Chapter 13

I f you wish to build a new dwelling or structure, create an extension, modify an existing dwelling, or convert a barn into a dwelling, you will need to deal with your local authority with regard to gaining suitable planning permission. You will most likely also have to make a building regulations application either to the local authority or to an approved independent inspector.

Complying with these requirements will ensure that you build a structure that is suitable for the area in which it is built,

healthy and safe for those who live in and around it, energy-efficient, and that will be accessible for people of all abilities.

Although both these regulatory bodies can seem confusing and restrictive, put there simply to control and annoy, if properly understood they can help you greatly in your design and building process.

In this chapter we hope to demystify both, helping you to understand them specifically in relation to structures that are made out of cob (and other natural materials).

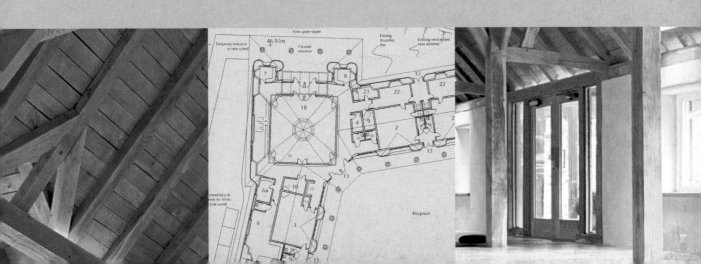

Planning Permission

Planning permission and building regulations are dealt with by two separate processes, and involve two very different sets of regulations. Planning permission literally involves gaining legal permission from the local authority to build a new structure, extend or modify an existing one, and other changes of use.

Regional planning policy

All local authorities have a defined planning policy, which determines a designated and specific land-use zoning for all development. Most zoning categories fall within the following: land assigned for agricultural purposes, green-field sites (land not previously built on or developed), green-belt (an area of parks, farmland, etc. encircling an urban area, with restricted development rights), brown-field sites (where industrial or residential buildings have previously existed and have since been demolished), conservation areas (where historical buildings exist and therefore have restrictions on new building), or designated new residential development areas.

Because the UK is such a small and crowded island, this zoning, although apparently controlling and restrictive, is necessary to avoid people building whatever they want, wherever they want. It is a way to manage development and essentially to protect the environment in which we all live. The planners, when making a decision,

work from a set of guidelines directly relating to the specific planning policy of their local authority. Thus when an application is made to the planning department they are concerned with three major concerns:

- What the designated land use is for the area in which the building is intended to be built, and whether it falls within the designated development boundaries.
- What the building is intended to be used for.
- What the proposed building looks like.

Intended building use and designated land use

The first and second concerns deal with the intended building use – whether it is intended to be a dwelling, a restaurant, a workshop, a holiday cottage etc. These will all have ramifications for what levels of access are required, whether there are municipal means to deal with waste water and sewage etc., or whether, for example, a wood workshop would create too much noise for existing residential neighbours. Although the rules are not hard and fast, you will generally not be able to step outside of designated land use categories described earlier. Although this may be frustrating on one level – if you want to build a humble, cob house and small-holding deep in the countryside – it is very reassuring and practical on another level, because it will (hopefully) prevent someone from building an obnoxious luxury mansion in the middle of a peaceful wheat field.

With respect to changing the use of a building, such as from an agricultural use (a barn) to a dwelling, the decision will generally be made within guidelines that relate to the bigger political picture. For example, where we live in Cornwall there has been a great emphasis on promoting tourism as a way to boost the lagging economy. This means that barn conversions into dwellings for holiday lets have been given very freely, whereas it is more difficult to gain permission to make a permanent dwelling. Hopefully things are changing.

Full planning permission is needed before any building work begins. If you do not have this, and your work is discovered, the legal authorities have the power to require you to return the site to its original state. It is a great idea to arrange to have a pre-application meeting, preferably on-site, with a local planning officer. This will start the fostering of a positive personal relationship with the officer, who should be able to give you constructive feedback on your project to assist you in making a successful application. In the long run, it may prevent you from receiving an outright rejection to your plans, because the officer can pre-empt potential and simple problems such as restrictions on moving natural features of the land such as hedgerows, on window orientation, or access. A few tweaks to your plans pre-application may mean the difference between success and failure, a battle or a partnership.

Ultimately, the decision for full planning permission is made by the council planning committee, which consists of local elected councillors who are advised by the planning officer. If you have a good relationship with your planning officer, he may look more favourably on your case, even if it does not fit perfectly into their guidelines. Local residents also have an opportunity to put forward objections via a public posting of the proposed development on the site.

For these reasons, it is best to be as transparent as possible with your project, both to your planning officer, the local council, and your neighbours. If people understand what you are building, and how you are building it, they are far more likely to support you. Cob can be a wonderful tool for building bridges in this way – you could organise an open day to get the locals involved (who can resist an opportunity to get muddy and stomp cob!). This way people can begin to appreciate how low-impact the process and the final outcome of building with cob is, and how unimposing your building will be.

Building aesthetics

The third concern deals literally with what you hope your building is going to look like. Although this is a purely subjective issue, planning policy generally looks favourably on buildings that blend in with the existing environment and traditional housing stock of the area. So, if you live somewhere with a tradition of earth buildings and thatched roofs, the planners should have no problem with the aesthetics of a new cob and thatch structure. In any event, cob buildings are so gentle on the eye, and blend so seamlessly into the British landscape, that you should never have a problem convincing a planner on this point, no matter how unusual your design!

Do I need planning permission for a small cob structure?

Planning permission may be needed even if you just plan to build a small cob structure in your garden. Every region is different, and specific advice will need to be sought from your local planning department, but generally, if your property already has planning permission, you should have rights under 'Permitted development' (check you have this). If so, you should be able to add a conservatory, porch, garage, or separate structure such as a studio, granny annexe or summer house, without having to go through the full planning process. If you build a separate structure, there are some guidelines that will need to be observed. To gain more information about this you can get a leaflet on 'Permitted development' from your local planning department, or go to your local county council website.

Self-build, the housing crisis and Section 106

With the current housing crisis in the UK, it is hoped that planning policy will change to encourage individuals and communities to take it into their own hands to self-build their own dwellings – for which cob is the perfect material. In the meantime, working co-operatively within the system can get good results. For example, in some regions where there is a perceived need for local housing (e.g. in many rural areas) permission for a dwelling may be granted under Section 106, which is generally an agricultural tie, even if it falls outside the local planning policy. The Section 106 comes attached with some stipulations, which, although different from region to region, may include financial ties to the council when the building is sold off, or the council having control over who can qualify to live in the building.

Agenda 21

Additionally, with the growing awareness of sustainability and environmental issues in governmental bodies, Agenda 21 has been developed as part of a European directive. This encourages, among other things, building methods and materials that are less harmful on the environment and that come from renewable sources. Consequently, if you use cob and thatch and other low-impact materials to build your house, and can incorporate a low-energy space and water heating system, the planners may look more favourably on your proposed project even if it falls foul in other areas. Bear in mind that some local authorities are more enlightened than others.

In conclusion, it is helpful to understand why planning policy is there and what specific policies are in place in your area, and don't let them frustrate you. Allow your planning officer to educate you about your site and region, and he or she will then more likely be open to you educating him or her about your building methods and materials.

For more information on the nitty-gritty of applying for planning permission, you can acquire a pamphlet called *Planning: A Guide For Householders* put out by the Office of the Deputy Prime Minister (ODPM) from The Stationery Office (TSO).

Building regulations

The purpose of the building regulations is to ensure that the building is reasonably safe and healthy for those in and about it, that the building complies with a minimum standard of energy efficiency with regards to its level of carbon emissions, and to ensure that reasonable access is possible for people of all abilities.

If your building has a floor area of less than 30 square metres and is not for residential use, such as an art studio or a garden shed, it will not need to comply with building regulations. If your building is a place where the general public will be entering, or is for full residential use, it will need to comply, whatever its size. This relates to all new buildings, extended and modified buildings, and some structural modifications such as widening existing openings, creating new ones, or removing walls or chimneys.

There are two types of building regulation bodies that can legally assess your building, and grant you a regulation compliance

certificate. One body is connected to the local authority, and the other involves hiring a private sector building control engineer known as an approved inspector. The latter is completely independent from the local authority, answerable to the Deputy Prime Minister, and regulated by the Construction Industry Council (CIC). Collectively, local authority building control and approved inspectors are known as Building Control Bodies (BCBs). It is your choice which one you choose. Using an independent inspector has benefits that will be discussed later in this chapter.

In contrast to planning permission, which is subjective and changes with political and departmental policy changes, building regulations should be objective (although they are open to interpretation and professional judgement), as they are a science, concerned primarily with the scientific engineering of a building – essentially, whether the building is constructed out of sound materials, in a sound manner.

Similar to planning officers, building control officers have a stigma attached to them for being notoriously difficult and contentious – especially when it comes to building structures that may be slightly alternative to the standard, modern, predictable, concrete block house. However, things are changing, and even though there may be a few dinosaurs around, whether you decide to hire an independent approved inspector or go through the local authority channels, most inspectors are incredibly knowledgeable, open and friendly, and can genuinely help

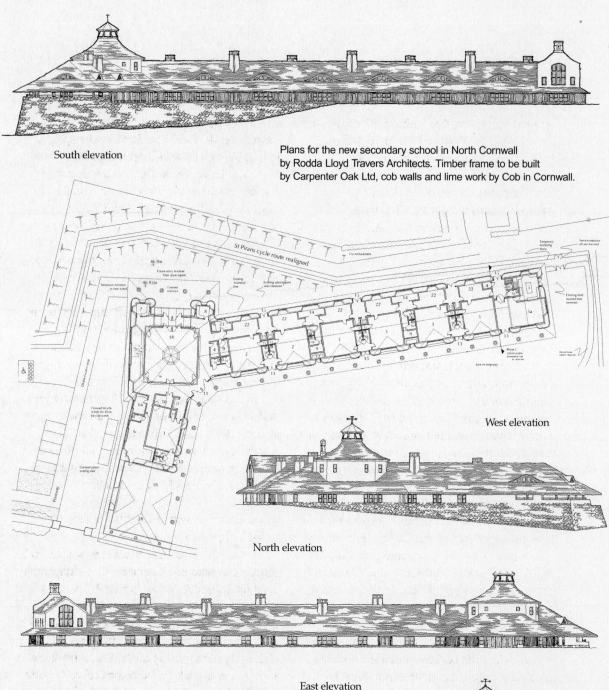

South elevation

Plans for the new secondary school in North Cornwall
by Rodda Lloyd Travers Architects. Timber frame to be built
by Carpenter Oak Ltd, cob walls and lime work by Cob in Cornwall.

West elevation

North elevation

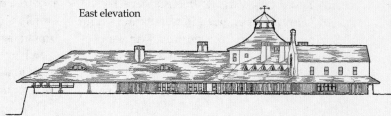

East elevation

This eco-friendly design
harmoniously integrates into
the rural landscape and makes
extensive use of cob walls.

you to design a building that meets all of the above-explained criteria – which will be better for you, and better for the planet.

The history of cob buildings and the building regulations

To understand why new cob buildings have a reputation for being notoriously difficult to get to comply with modern building regulations, it is necessary to look back on the historical development of these regulations, and the circumstances in which they developed. Prior to 1666 and the Great Fire of London, there were little or no official building regulation standards. The Great Fire of London, which started in a bakery, created so much devastation because most of London's buildings were constructed out of wood. The consequent planning act therefore stipulated that buildings should be constructed primarily out of non-combustible materials such as stone and brick. There was an absence of cob buildings in London because the clay was unsuitable for building, and consequently, from day one, cob never made it into the policy documents, despite the fact that probably half the buildings of the country were made out of it!

In subsequent years, the policies were changed and expanded upon, but consistently failed to include buildings made from cob. This was largely due to the fact that most cob buildings (though not all), were built in rural areas, by the rural workforce, which were largely out of the jurisdiction of national policies. In 1965 the building regulations policy determined a 'fitness of materials'

specification, which set out a particular British Standard Code of Practice for specific materials such as stone, steel, cement, and brick. Again, cob was not included in this line-up, because it had no British standard (probably because it was so difficult to achieve a predictable standard with such a varying material). With no standard, it could not comply with the regulations, so new cob buildings all but disappeared from the construction scene. All of this changed in 1985, when a window of flexibility opened up with a change from 'prescriptive regulations' (that is, regulations attached to specific rules and instructions), to 'functional' or 'performance' requirements. This means, that as long as you get to the requirement in the end – for example, building a wall that is structurally sound – you can get there how you like. From this moment, by default, cob was once more eligible as a *bona fide* material with which to build a new structure.

Cob and the modern building regulations system

This new system involves a series of technical requirements to be complied with, along with a set of Guidance Notes, labelled A-P. These cover all aspects of providing safe constructions, with regard to items such as sound insulation, structural stability, drainage, waste disposal etc. There are some requirements that can be complied with quite simply, such as part F1, which specifies that you must have adequate means of ventilation, but some which are much more open to interpretation, such as Regulation Seven regarding materials and workmanship, as the Guidance Notes state: "Any material which can be shown by experience such as a building in use, to be capable of performing the function for which it is intended" is satisfactory. In this respect, cob could be interpreted as a highly suitable building material for a new construction.

Out of these technical approved documents, there are five which have provided challenges for the passing of a new cob structure in the past. This is because either there is no data available to prove or disprove its viability in certain areas, or because it does not quite reach the standards imposed by the modern regulations. Nevertheless, with a little extra thought, cob can prove itself to be a more than suitable, modern, building material. These five points are:

- Regulation 7 – Materials and workmanship
- Part A – Structure and stability
- Part B – Protection from fire
- Part C – Site preparation and resistance to moisture
- Part L – Conservation of fuel and power

Regulation 7 – Materials and workmanship

Regulation 7 is designed to ensure that all building work is executed with appropriate materials in the appropriate circumstances, and that they are put together with proper 'workmanship'. It also considers the environmental impact of building work and materials, as stated in the approved guidance document: "The environmental impact of building work can be minimised by careful choice of materials, and where appropriate the use of recycled and recyclable materials should be considered. The use of such materials must not have any adverse implications for the health and safety standards of the building work."

It needs no further explanation as to why cob falls perfectly in line with this recommendation, and as long as the cob mixture used is not contaminated in any way, it will provide no problems with regards to the health and safety standards of the building work. We would also recommend, if using recycled cob from an old building, that the mix has an adequate amount of clay remaining (the clay element could have been largely washed out if exposed to the elements for a long period of time), and that there is still enough fibrous material remaining. This element could have disintegrated. Best practice when using old cob is to mix fresh straw in with the batch. This will ensure that the cob has optimum compressive and tensile strength.

"Proper materials" can be deemed as those that have either been approved by testing bodies such as the British Board of Agrèment (BBA), or by tests carried out by other independent accredited laboratories such as at Plymouth University's Centre for Earthen Architecture, who have developed a series of tests to establish a performance specification for soils, to determine their suitability for building, or the Department of Architecture and Civil Engineering at the University of Bath. It is also possible to use specifications and data from other EU countries when there is no data available in the UK. On the whole, other EU member states have been more advanced in their research and acceptance of traditional earth building techniques, the most notable being the International Centre for Earth Construction (CRATerre) in France, and the University of Kassel in Germany. "Proper materials" can also be deemed as those that have been proven by successful practice.

As mentioned earlier, Regulation 7 states that past experience of buildings constructed in the same manner, as long as they have stood the test of time, can help them to be deemed viable. This is very relevant for cob, as many parts of the country have examples of functioning ancient cob houses, which can be used to prove the material's viability in a new structure. If the new cob structure is to be situated in a region with a history of cob building, reason dictates that a local subsoil will be suitable. If wanting to build in an area where there is no such history, it is common sense that soil testing will need to be carried out to determine its suitability, which any competent builder should do anyway. Each building control body may react differently – they may be satisfied with positive demonstrations of your own tests as outlined in Chapter 3 (go for the jar test, and not the sausage test!) They may want to see the soils tested in a credible laboratory such as at the University of Plymouth or the University of Bath, or they may take samples of the material and do their own tests.

With regards to "workmanship", as there is

no standard relating to cob construction, it is entirely possible to comply with this document as long as you can demonstrate that you have a good understanding of the material, and have some practical experience of using it. Proof of attending a course should be adequate. You may even know more than your building inspector (unless you live in Devon!).

It is thus very clear that as long as Building Control can be assured that properly mixed, suitable soils are used, along with good building practices, cob can comply with this requirement. And of course you should want nothing less from yourself, or your hired professionals, for your own safety and peace of mind.

Part A: Structure and stability

Part A requires that the dead load of the building (i.e. the weight of the building) and imposed load (wind and snow) be transmitted to the ground safely. Essentially, your officer will need to be convinced that your structure is not going to subside, slide down a hill or collapse under the load of the roof.

Assessing your soils

For this, the suitability of the subsoil will have to be assessed, to prove that it is good for building on. Your building inspector will be able to help you assess this, or a structural engineer's report can be sought.

From these investigations a suitable foundation design can be implemented, and as described in Chapter 5, should be created

300mm/12" (i.e. 150mm/6" either side) wider than the width of the cob wall and the stone plinth. This is to spread the very heavy load from the cob walls, floors and roof.

Cob as a load-bearing material

Your building inspector will want to know that your walls are strong and can take all loads. A well-made cob mix, made with good materials and well compacted, will have excellent strength in compression. As a load-bearing material, therefore, cob is excellent, but it must be built thick to do this effectively. This is why you will generally never see a structural cob wall built under two feet thick; this is so that the forces can pass down through this thickness into the ground.

Tests have been done by Linda Watson and Bob Saxton at the University of Plymouth to show the average density of a cob wall. This ranges from 1,700–1,900 kg/m^3 (higher levels of coarse aggregates provide a higher density), and the density has an impact on its strength. The higher the density, the stronger it is, with an average of 0.77 N/mm^2 (111 lbs per sq. inch). Figures provided by the Devon Earth Building Association (Structural Survey, vol. 15, no. 1, 1997, pp. 42-49) show that cob can meet the loading and stress requirements for Part A: Structure and stability. For example, a 450mm (18") cob wall with a 750mm (30") wide foundation for a two-storey building will have roughly the following profile:

- Average stress in wall of 0.16 N/mm^2 (23 lbs/sq. inch)

- Loading at base of the wall of approx: 68 KN/M^2 (9.8 lbs/sq. inch)
- Ground-bearing pressure of 90 KN/M^2 (13 lbs/sq. inch)

Other points to consider for structural stability

The structural stability of a cob wall is also dependent on a number of other factors, including the correct use of lintels over door and window openings so that the wall and roof loads above can be carried (external lintels must always be made out of good quality hardwoods); the correct attachment of the roof to the cob wall using a properly anchored and fitted wall plate; correctly fitted floor joists on the second storey of the building; the pitch of the roof, amount of openings, the weight of floors, etc.

If you live in an area where there is a tradition of cob building, you will be able to use these examples to support the evidence of the structural stability of cob made from the local subsoils. These points all relate to common-sense siting and design practices, which should be executed for the construction of any building, built of any material, at the very earliest stages of choosing a site on which to build, so that it can be made safe and durable for those living in and around it.

Part B: Protection from fire

Cob has good fire-resistance properties, and can generally be described as being 'non-combustible', despite having organic fibres present (straw). Research done in Germany by Gernot Minke (2000) at the

University of Kassel has proved that earthen walls are non-combustible as long as their density is no less than 1,700 kg/m^3. This data has been recognised as a standard in German building codes.

Past collapses of cob walls during fires are mainly due to their being soaked by water as the fire is put out, which can cause the cob to become plastic and lose its form, and not because of their inability to resist the fire. It can also be due to the collapse and failure of burning structural timbers such as roof trusses and floor joists. The standard application of two coats of lime on the inside and outside of the cob walls will further increase this fire-resistance.

Part C – Site preparation and resistance to moisture

Part C deals with the preparation of the building site and protecting the building from the damaging effects of too much moisture penetration into its fabric. The former is concerned simply with the correct preparation of the site before building commences, and pertains to the same practices that would be upheld in a conventional build. Where special consideration should be taken with a cob build, is when the subsoil to make the cob is to be extracted directly from the site. In this instance it must be proven that the subsoil is not contaminated.

The issue of resistance to moisture relates to "The ability of all parts of a building to resist the passage of moisture from the outside to the inside of a building" (Approved Document guideline). Too much moisture in a building, or at least moisture that is trapped, is unhealthy for its occupants, as it can cause toxic moulds, fungi and respiratory problems. Additionally, for cob walls, excessive contact with moisture has the potential to cause failure of the wall.

For cob walls to be free of damp and to be protected from the damaging effects of excessive moisture ingress, three areas need special attention: the application of permeable, breathable finishes such as lime plasters/renders and washes, instead of the conventional, impermeable cement-based ones; the installation of an effective drainage system that will take water rapidly away from the building; and the construction of the necessary stone plinth built up from the ground, onto which the cob is laid, to prevent it from coming into direct contact with moisture from the ground.

Impermeable, breathable finishes

As discussed in Chapter 9, cob's potential for creating a damp internal environment is due mainly to the unsuitable application of non-breathable finishes such as cementitious renders and acrylic paints, which can lead to the trapping of moisture at the base of the wall, which in turn can cause damage and problems with damp inside. The use of lime and other breathable finishes will create a healthy exchange of moisture vapour, healthy walls, and a healthy house.

It is therefore important to understand that the moisture balance mechanisms of a cob structure work differently from those of a modern concrete-block house. The former depend on the lime to act as a sacrificial layer, absorbing the moisture like blotting paper and then releasing it as the air wicks the moisture away and the building dries. The latter rely on impermeable barriers such as cement renders to prevent moisture from entering any part of the fabric of the building. Your building inspector should be aware of how cob buildings behave differently, to enable the above rationale to be accepted as compliance with the regulations. With a breathable cob structure, it is not so much a matter of 'resisting' moisture, but of using methods that allow the building to reach its own natural equilibrium.

Effective drainage

Common-sense drainage systems are essential to take water away from the building as quickly as possible. This may involve simple self-draining foundations or land drains, which will negate the need for a plastic damp-proof course (DPC). Plastic DPCs are not recommended for cob walls because, like concrete, they are largely impermeable and non-breathable, and may cause moisture to accumulate at the base of the walls. For more information on this, please refer to Chapter 5: *Foundations*.

Stone plinth

The traditional and necessary stone plinth, built to a minimum height of 600mm (2'), serves to provide a capillary break between the ground and the cob, and therefore avoids the risk of moisture rising into the

base of the cob wall. Again, information on this can be found in Chapter 5: *Foundations*.

The design of a cob building with a thatch or turf roof without guttering must also take into consideration the overhang of the roof eaves. These must be at least 400mm (16"), so that precipitation falling off the roof is thrown clear from the walls. Additionally, the drainage system must be able to take this moisture quickly away from the building.

Part L: Conservation of fuel and power

This requirement relates to how energy-efficient and thermally efficient a building is, and states that a building should have elements within it that go some way to limiting heat loss through the fabric of the building, hot water vessels and pipework. It also includes the control of the space heating and hot water systems through thermostats, and the efficiency of the lighting. The issue of the thermal efficiency of a building (i.e. how easily the materials making up the building lose heat from the inside to the outside) were first raised in the 1950s, to ensure a level of comfort and well-being for the occupants of the building. In the 1970s, this regulation was extended to cover the conservation of fuel and power in response to the oil crisis happening at that time (household energy use comprises around 53% of total energy use). The contemporary relevance of this requirement is in response to the global warming crisis, brought about by the 'greenhouse effect', and the need to reduce CO_2 and other greenhouse gas emissions.

Re-built cob carthouse, Cornwall, designed by Mathew Robinson, cob walls by Cob in Cornwall, thatch roof by Mike Pawluk, timber frame by Stefan Roux, stonework by Paul Finbow.

" If you want to do something new, you should find the questions asked by Building Control helpful, because if you can't answer them, you haven't thought about things enough. "

For modern buildings, therefore, the regulations are concerned with the actual emissions created by the building over its lifetime, such as the oil burnt to fire the radiators, the gas to heat the water, the electricity to light the building, etc. It is clear that a building that is very well insulated, which contains a sustainable heat source, and well-insulated water pipes, and that is oriented to maximise passive solar energy gains will use less energy than a building that does not contain these elements.

Unfortunately, as it stands right now, the regulations do not take into account the emissions created by the extraction, processing, transportation, construction and eventual disposal of the materials used to construct the actual building. However, seeds are being sown in this area which should bring about proper accounting for these valuable areas of energy expenditure (called 'embodied energy') in the not too distant future. It is clear that a building constructed out of cob and other local, minimally processed materials, would perfectly fit the bill with regard to the conservation of fuel and power. A cob building that uses a renewable-energy system for its water and space heating, such as a wood pellet burner photovoltaic system or windmill, would further decrease the amount of energy consumed and emitted. It can also be sensitively designed to utilise the principles of passive solar – capturing the energy of the sun to heat and light the building. A cob building constructed in this way could be said to be almost the perfect 'eco-build'.

Where cob has traditionally been perceived to have shortcomings, and has therefore had challenges in being able to meet the building regulation requirements, is in regard to its inherent thermal efficiency, or insulative properties. This relates to the U-values mentioned in Chapter 8: *Insulation*, which measure the amount of heat a material allows to pass through it.

A cob wall 650mm (25") thick has a U-value of roughly $0.66W/m^2K$. With render, this drops down to $0.55W/m^2K$ (the denser the wall, the higher the U-value). Previous stumbling blocks around meeting this requirement have occurred because each elemental part of the building has been required to meet the specified U-value – for example the roof, the walls and the floor. Calculated this way, a cob building could not meet the required standard, even when rendered and plastered inside and out and built on a well-insulated stone plinth.

However, new legislation has done away with this system of calculating the thermal efficiency of a building, and the new system takes a more holistic approach, which takes into account the efficiency of the whole building and not just singular elements within it. This is where there is new hope for buildings constructed out of cob.

A holistic approach: target U-values, and whole building energy efficiency

It is now possible for a new cob building to fully comply with the building regulation requirements using this holistic approach – a fitting method of analysis for a completely holistic building system. Using this approach, the thermal efficiency of a building can be analysed by calculating the cumulative U-values of all elements of the building together, allowing for the high U-values of the walls to be compensated for by extra efficient window glazing, and roof and floor insulation, as long as no element falls below a certain target U-value (0.7W/m²K for walls).

A cob wall which is 650mm (25") thick will have a U-value of roughly 0.66W/m²K, and when rendered with lime inside and out will have a U-value of roughly 0.55W/m²K. This means that a sensitively and intelligently designed cob building, which also incorporates renewable energy heating systems and passive solar gains, and utilises internal heat gains through its excellent thermal mass properties, can easily meet the requirements set out in Part L.

Energy-efficiency design elements for cob structures

To ensure that you construct a highly energy-efficient cob building, and adequately compensate for the high U-values of a cob wall, we recommend that you consider the following design and construction details:

1 Insulate as much as you can all other elements of the building: the roof (even a thatch or turf roof), the floor and the foundation plinth. Refer to the relevant chapters for specific guidelines.

2 Render the cob on the outside and plaster the cob on the inside (with a breathable material – lime is ideal). Experiments have been done by the University of Bath on the effect of applying a lime-based external render which contains pumice stone aggregate, onto earthen walls. This has the potential to increase the thermal efficiency of the wall without interfering with breathability.

3 Buy the best quality glazing you can afford for all glazed openings: the minimum standard should be a low–E double-glazed unit. The most effective are sealed double or triple units with wooden frames, which have better insulative properties than plastic or metal frames.

4 Use insulation shutters in winter on large expanses of glass, and thermal curtains on other window openings.

5 When fitting lintels, always use separate pieces of timber to span the width of the wall, rather than one piece (which will provide a bridge for the cold of the outside to travel to the inside, and the warm out).

6 Consider insulating the inside or outside face of a cob wall using lightweight materials such as wood, insulation, tiles, or breathable, environmentally friendly alternatives to plasterboards, such as Pavatherm.

For example, you can externally face the cob with timber cladding or shakes or shingles, with an air gap between the timber and the wall; or internally attach a breathable plaster board such as Pavatherm to a timber framework, with an insulated cavity in between, with a lime plaster finish. Bear in mind that lining the internal face of a cob wall will interfere with its thermal mass properties from internal heat gains, so this option could be limited to the coldest parts of the house, such as the north-facing side.

7 Consider constructing the north, 'cold' wall out of straw bales, which have excellent insulative properties.

8 If you construct a small house, build a floor to ceiling bookshelf on the north, cold wall, and insulate between the back of the bookshelf and the wall with a breathable material such as sheep's-wool. The books should help insulate too!

9 Install a renewable-energy space and water heating system to reduce your carbon emissions. The new Part L takes this into consideration, and provides a method to calculate exactly how much you will need to reduce your carbon emissions in relation to the overall energy efficiency of the building.

10 Site and design your building to protect from fierce weather, and employ passive solar principles to capture the sun, so that you can exploit the excellent thermal mass properties of cob. Orientate your building so that its longest face is facing south, concentrate most windows on this side, and minimise windows on the north-facing wall.

11 Incorporate a sun space or conservatory onto the east or south walls of the building, to create a buffer zone for cold air entering into the building and, again, to exploit the thermal mass properties of cob.

12 Use a lot of straw in your cob mix, as this will improve its insulative properties (the hollow stems will hold air); and though an expensive, non-local option, consider adding insulative lightweight mineral aggregates such as vermiculite, perlite or expanded clay balls to your cob mix, again to improve the insulative properties of the cob. However, be aware that too much aggregate added in this way could interfere with the compressive strength of the cob, and hence its load-bearing capacities.

13 Where open flues and chimneys are present, ensure that they are fitted with a damper that can be shut off when not in use, to prevent the loss of heat from the building.

14 Make sure that all windows and doors are tightly sealed to prevent heat loss.

15 Consider building your insulated wall plinth higher than the recommended two feet, necessary for its function as a capillary break.

The possibilities are endless, and this list is not exhaustive! Some are common sense and general, others necessitate creative thinking. Do be aware however that, as shown in some of the suggestions listed above, the addition

(continued on page 236)

An interview with Jon Hollely, director of jhai Ltd. and Mark Saich, environmental scientist and sustainable construction manager for jhai Ltd.

jhai Ltd is a company offering full building regulation inspection and approval services as a direct alternative to the local authority. They actively support and promote sustainable building materials and methods of construction, and are able to offer expert advice on low-impact designs and materials. The following interview will illuminate exactly who they are, what they do, and why their services are potentially extremely beneficial for anyone wanting to construct, and get approval for, a new dwelling or the conversion of an old structure that is made out of cob.

Who are you and what do you do?
Who are you answerable to?

We are jhai Ltd. Our core business is as an approved inspector, that is, private sector building control, working for clients as an alternative to the local authority. We are leaders in building control for sustainable construction. I (Jon Hollely) was the technical officer for the straw bale association for three years. We are the only building control body in the country who have a qualified environmental scientist (Mark Saich) working full time. We are answerable to the Office of the Deputy Prime Minister via the Construction Industry Council – an independent body appointed as the Deputy Prime Minister's agents.

What are the benefits of using your company to pass an 'eco-build' through the regulations?

The benefits of using jhai Ltd to ensure that an eco-build complies with the building regulations are that many other building control bodies do not know much about sustainable construction technologies. jhai Ltd. can draw on a wide and in-depth knowledge of many aspects of alternative construction including technologies, materials, renewable energy sources, and waste management. Our company is committed to actively promoting sustainable construction.

Our approach is to partner with our clients, rather than just saying yes or no, to find appropriate solutions for their building scheme. We like to communicate right at the beginning to assist with the design process so that we can be pro-active rather than reactive to their designs, and can assist them towards appropriate solutions. Partnering and teamwork are our keywords, preferring to be pragmatic rather than dogmatic, whilst always ensuring compliance with the regulations.

On a practical level, we give an extremely fast response – issuing an approval in as little as two hours, compared with two months from the Local Authority. We pride ourselves with providing technical consistency and national uniformity.

What is the purpose of the building regulations today, especially with regards to Part L and the energy efficiency of the building?

The purpose of the building regulations today are to ensure minimum standards of health and safety to those in and around a building, to ensure good levels of energy conservation, and to ensure access and facilities for all people. These are achieved through complying with certain require-ments as set out in Schedule 1 of the Building Regulations, A–P. Along with each set of requirements comes an approved document which provides guidance as to one method of how to achieve compliance.

In 1985, a new system introduced functional requirements, which means that as long as the end requirement is achieved, you can reach it how you like. This is instead of prescriptive regulations, which stipulated very specific rules that had to be adhered to.

Part L – *Conservation of fuel and power*, is a very important part of the modern building regulations. They contain a package of measures and tools that the UK are developing to go toward fulfilling the Kyoto obligations to reduce the green house gases and substances which contribute to global warming.

Why is it difficult sometimes to pass a building through local authority channels, when the materials and methods used deviate from the standard, modern practices? For example, cob and lime, straw bale, etc.

There are many good professionals in local authority building control who have similar attitudes to us, but unfortunately, there are some who still act like old-fashioned inspectors, who can be 'closed minded'. They are an inevitable part of bureaucracies who have their own agenda. It is not uncommon for us to have approved an application in the time it would take a local authority to send an acknowledgment letter. This, combined with our knowledge and commitment to promoting sustainable construction, brings benefits to the client.

As mentioned before, we prefer to partner with our clients, and are open to learning from their expertise about a particular construction method. In the case of cob building, the proof is in the pudding. If we can see similar types of constructions in existence, that are working, we will use them to support the suitability of a new building constructed in a similar way.

In theory, our approach is no different from the best local authority inspectors. The schedule and method of inspection are exactly the same.

We (the authors) work with natural building materials, and our ethos is to minimise the use of pre-fabricated, high energy-consuming and toxic materials wherever possible. For example, we prefer to use a solid stone foundation plinth instead of the local authority standard of a stone-faced concrete block cavity wall. We like to avoid at all costs using a plastic damp-proof course or membrane because this is not suitable with cob. We like to avoid using a concrete slab for the foundations. Is it possible to avoid these and still bring the building into compliance with the building regulations?

Yes. We recognise that some of those things would have a detrimental effect on a cob building and create more problems than they would prevent. For example, installing a damp-proof course is not a requirement, but reasonable precautions taken to prevent damp are, such as good drainage and good ventilation. Incidentally, the cavity wall you refer to is a designer standard and not a local authority one.

Would you be happy to pass a new building with walls made out of two-feet thick cob, even though some local authority inspectors have been reticent in the past? How would one achieve this, and what measures can be taken to compensate for the high U-values of a cob wall?

Yes. We would take a holistic approach to the building, looking at the efficiency of the whole building. For this, we would use a target and design-efficiency rating, as opposed to looking at the efficiency of each individual element of the building. This includes looking at the energy efficiency of the heating system and insulating the cob if necessary.

To compensate for the high U-values of a cob wall, we would recommend that you super-insulate other parts of the building such as the roof, and that you think about designing the building so that it can benefit from passive solar heating, such as orienting the building towards the south, with glazing on this south-facing side.

What are the specific criteria for a new, modern thatch roof that is thermally efficient?

A new 300mm-thick thatched roof will provide adequate insulation for a modern, thermally efficient house. For a building made out of cob, it would be beneficial to install extra insulation in the roof cavity between the rafters to increase the overall thermal efficiency of the building.

of certain elements may improve one area of the building's thermal performance, only to compromise another. All trade-offs should be considered within the overall soundness of the building.

Energy efficiency assessment techniques

To assess and analyse the energy efficiency of a building using these holistic approaches, a number of assessment procedures can be employed. These will either provide an overall score or rating, or can provide a thorough set of details about how 'eco' your building and whole lifestyle are.

SAP

The most basic – the Standard Assessment Procedure (SAP) – produces a rating from 1 to 120 (the higher, the better). It is a specification for all newly-built houses, to ensure that they are designed to a standard energy rating. It will also calculate the annual energy costs of space and water heating. The results are displayed mandatorily for potential buyers, so that comparisons can be made. Along with the SAP comes a carbon index rating, which determines the annual CO_2 emissions of space and water heating systems in regard to the levels of pollution of the type of fuels used.

NHER

A more accurate energy rating can be sought through National Home Energy Rating (NHER), which is the same as the SAP but more accurate and sensitive. It is more thorough, and therefore more expensive than a SAP, and can only be carried out by a trained assessor. This can give you a SAP rating for building regulations to help you comply with the Part L requirements.

BREEAM

Even more comprehensive and holistic in its approach is the Building Research Establishment Environmental Assessment Method (BREEAM), which aims to produce a rating for domestic buildings with regard to a broad range of environmental issues such as ecology and land use, energy and water use, materials used, health and well-being, transport issues, and the pollution associated with the whole building, starting with the construction process. This sort of assessment could provide an excellent tool to be used to help a new cob building meet the requirements of Part L (Conservation of fuel and power).

The regulations in relation to old, existing cob structures

This refers to existing cob buildings where a change of use occurs, e.g. a cob barn under agricultural use that changes to a dwelling. Allowance is made for old buildings that are constructed in a way that would limit their ability to comply, but this is on the proviso that elements of the building that are reinstated, such as roofs and windows, are done so in a way that will help to counterbalance the 'limited' areas.

Extensions to existing cob buildings

All new work must comply with the regulations.

Conclusion: the benefit of the modern building regulations for cob

From the above information, it can be seen that the science of modern building control is being opened to much broader environmental issues, and being subjected to a wider set of concerns imposed by European and global initiatives. The questions it raises for the owners and builders of cob structures surrounding thermal and energy efficiency should be regarded as very positive, and seen as a stimulus for seeking out appropriate solutions. The modern building regulations, and the Part L documents (Conservation of fuel and power), have definitely not excluded cob from being considered as a viable, modern, 'green' building material. They have in fact provided a catalyst and a vehicle through which cob can be considered a highly valuable building material within the context of a holistically designed, energy-efficient structure, where the whole is greater than the sum of its parts. This example is a true illustration of how we can look at past technologies, and adapt and improve them to suit our needs of the present, so that they can be considered modern, responsive and highly relevant.

Your relationship with your building control officer can be a dynamic one. Use him or her as an impartial pair of eyes both at the design stage and as work progresses. For the innovative self-builder, this can be hugely advantageous, and should be seen as an opportunity to confirm the efficacy of your work.

Jon Hollely (see interview on pages 233–5) raises some interesting points for building control compliance in relation to new or 'alternative' building methods (for although not new, new cob construction can sometimes be misconstrued as being whacky or hippy) by stating that: "If you want to do something new, you should find the questions asked by building control helpful, because if you can't answer them, you haven't thought about things enough."

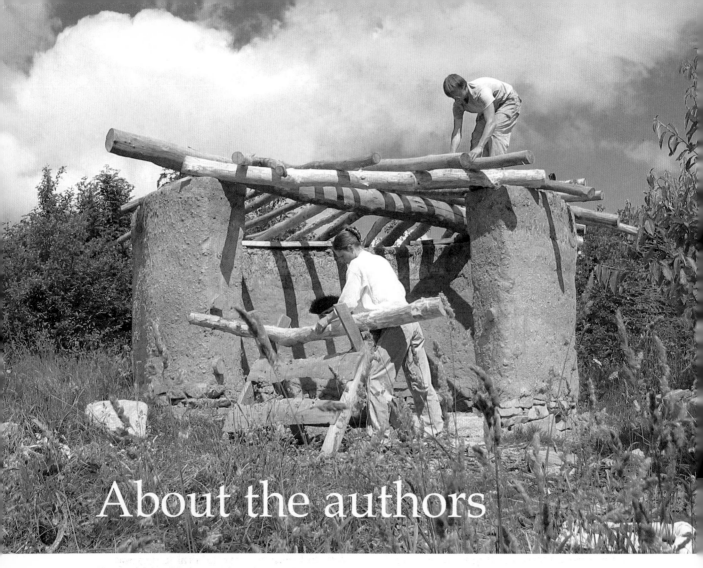

About the authors

Katy and Adam jointly at work on their garden retreat/studio.

Adam Weismann and Katy Bryce undertook an apprenticeship in natural building with the Cob Cottage Company, Oregon, USA. They then returned to the UK to work with Matthew Robinson for a short while, before starting their company Cob in Cornwall (www.cobincornwall.com).

Cob in Cornwall is based in Manaccan, on the Lizard Peninsula of Cornwall. It is a small company specialising in the new build and restoration of earthen structures, known as 'cob' in the south-west of England.

They approach all of their projects with an ecological ethos. Katy and Adam have been and are actively involved in the restoration of many ancient cob buildings – repairing walls using pre-dried cob blocks, and applying appropriate 'breathable' finishes in lime.

Adam and Katy also build new cob structures, including schools, two-storey houses, small studios and garden rooms, outside courtyards and fireplaces, and earth ovens. In 2006 they are building an entire school out of cob in Newquay, Cornwall. They feel it is exciting to be using this ancient and simple technology in a contemporary setting.

Adam's Story

Adam Weismann was born in Iowa, USA, and came to England in 1999 to complete his studies in documentary film-making. It was then that he fell in love with his now wife and business partner, Katy. It was during a trip together to the West Coast of America, having left the big smoke of London to seek a more wholesome existence, that they discovered and embraced the wonderful art of cob building.

Adam loves working outdoors and using his hands to create, which he has been doing since he was a young boy. Early ventures include the construction of backyard neighbourhood skateboard ramps, and later a progression into working with stone and wood. All this meant that making houses and structures out of mud was an easy love affair to begin and carry on into the future.

Katy's Story

Katy was born in Welwyn Garden City, England, and obviously decided in her early years that she was a true garden baby. Her formative days were spent roaming around in green spaces and being fascinated with what came out of the earth. A degree in human geography, with a focus on the notions of shelter, especially in poverty-stricken areas, meant that her introduction to cob excitingly brought together two areas

that she felt very passionate about. Her desire to create beautiful, functional and sustainable shelter out of the earth began.

In 2003, Katy and Adam won a 'Pioneers to the Nation' award from the Queen for their work with cob. This confirmed that their company was not just a mud pie in the sky idea, but was firmly rooted in a growing public desire for dwellings and structures that are made out of sustainable materials, that don't pollute and destroy the planet, and that people can self-build.

In 2006, Katy and Adam are planning to build a house for themselves out of their favourite materials – stone, mud, wood and grass. They hope to create a building that contains a blend of the traditional and contemporary.

They are affiliated to Seven Generations Natural Builders, a natural building group who work in the Pacific Islands and the Pacific Northwest, USA. They come together to work on international projects and workshops. With them, they have established a non-profit organisation called 'By our houses you will know us'. The aims of this organisation are to foster an understanding of, and generate the education of vernacular building practices around the world, through the medium of sharing knowledge and reciprocal learning (www.byourhouses.com).

When not building, they enjoy hiking in the outdoors, swimming in the sea and growing vegetables.

Resources & suppliers

MATERIALS SUPPLIERS

Anglia Lime Company (Suffolk)
01787 313 974
Sabion render guns.

Carpenter Oak (Devon)
Elliot Atkinson
01803 732900
enquiries@carpenteroak.com
www.carpenteroak.com
Oak (green or seasoned) and other traditional building products such as hand cleft pegs for timber framing, plaster lath in oak or sweet chestnut, oak roofing shingles.

The Cob Block Company (Somerset)
Sam Owen
01823 673360
The-cob-block-company
@appley7028.fsnet.co.uk
www.thecobblockcompany.co.uk
Cob blocks.

Cornish Lime Company (Cornwall)
Phil Brown
01208 79779
phil@thecornishlimecompany.co.uk
www.thecornishlimecompany.co.uk
High quality lime putty, hydraulic limes and limewashes, pozzolans, lath, hair and fibres, sands, plastering tools, Aglaia natural paints, natural pigments, cob blocks, ready mixed lime mixes etc.

Delabole Slate Company Ltd. (Cornwall)
01840 212242
Delabole slate.

Dick Fine Tools (Germany)
+49 991 910930
info@dick.biz
www.dick.biz
A full range of traditional building tools and materials such as casein powder, borax, beeswax etc.

Earth and Reed Ltd. (Suffolk)
Christopher North
01449 722255
info@earth-and-reed.co.uk
www.earth-and-reed.co.uk
Natural building products and environmentally responsible building and decorating materials.

Energy Ways (Hertfordshire)
Rob Street
01920 821 069
enquiries@naturalinsulations.co.uk
Natural insulation.

Ewan Clitherow-Barker (Wales)
01654 781 354
Quarter-sawn oak roof shingles.

Firestone Building Products Ltd.
01423 520878
Rubber guard EPDM roofing membrane, ideal for turf roofs.

Greaves Welsh Slate Company (Gwynedd, Wales)
01766 830 522
Supplier of Welsh slate.

Green Building Store (West Yorkshire)
01484 854898
sales@greenbuildingstore.co.uk
www.greenbuildingstore.co.uk
Natural insulation, natural paints and high-performance windows and doors with super-efficient glazing.

Helifix Ltd
0208 735 5222
sales@helifix.co.uk
www.helifix.co.uk
Manufacturers of Helifix ties and attachments.

ICB Ltd (ALWITRA) (Dorset)
Bob Dixon – 01202 579 208
EPDM single-ply roof membranes, ideal for turf roofs.

J & J Sharpe (Okehampton, Devon)
Jerry and Jan Sharpe
01805 603587
mail@jjsharpe.co.uk
www.jjsharpe.co.uk
Specialist suppliers for the repair and conservation of old buildings, Helifix ties and Helifix driver, etc.

Old House Store (Oxfordshire)
Kate Dukes
0118 969 7711 kate@ijp.co.uk
www.oldhousestore.co.uk
*Traditional and ecological building
products.*

Narrow Water Lime Service
(Ireland, County Down)
016937 53073

Rawnsley Woodland
Products (Cornwall)
Tino Rawnsley 01208 813490
tino@cornishwoodland.co.uk
www.cornishwoodland.co.uk
*Sustainably harvested timber products
such as planed cladding profiles and
shingles, all from locally sourced red
cedar. All products are fully FSC-
certified and dried in a solar kiln.*

Safeguard
01403 210204
www.safeguardchem.com
*The Oldroyd XV water-proofing system,
ideal for turf/green roofs.*

The Scottish Lime
Centre Trust (Fife)
01383 872722
High quality lime products.

Second Nature UK Ltd.
(Cumbria)
Penny Randell
01768 486285
info@secondnature.co.uk
www.seondnature.co.uk
*UK distributor for Thermafleece
sheep's wool insulation.*

Sheep Wool Insulation Ltd.
(Wicklow)
David Pearce
+353 404 46100
info@sheepwoolinsulation.ie
www.sheepwoolinsulation.ie
Sheep's wool insulation.

The Traditional Lime Company
(Gloucester)
01242 525444
Suppliers of high quality lime products.

Ty Mawr Lime Ltd.
(Brecon, Powys)
Nigel Gervis
01874 658000
tymawr@lime.co.uk
www.lime.org.uk

Woodland Crafts Supplies Ltd.
(Suffolk)
01394 274419
woodland@fxferry.demon.co.uk
www.woodlandcraftand
supplies.co.uk
*All different types of beautiful
woodworking tools.*

Mike Wye and Associates (Devon)
Mike Wye
01409 281644
info@mikewye.co.uk
www.mikewye.co.uk
*Quality lime putty, hydraulic limes and
limewashes, ready mixed lime, cob
blocks, natural insulation, claytec
plasters, lath, hair, fibres, Helifix ties and
Helifix driver, timber products, etc.*

CRAFTSPEOPLE mentioned or featured in this book

Carpenter Oak Ltd (see page 241)

Green Oak Carpentry Co.
Andrew Holloway
01730 892 049
andrew@greenoakcarpentry.co.uk
www.greenoakcarpentry.co.uk

Paul Finbow, Stonemason
(Cornwall)
07967 103131

Stephan Roux and Co.
Timberframes (Cornwall)
01326 561 297 or 07812 561 297

Mike Pawluk
Master Thatcher
(Cornwall)
01209 831 804
mikepawluk@supanet.com
www.mikepawluk.supanet.com

Jamie Lovekin Timberframes
(Cornwall)
01736 763676

D'Ocres – Lime and pigment
specialists in France (St Antonin)
Bruno Mohamed and Preville
0563 56 13 03
contact@docres.fr
www.docres.fr

COB BUILDERS

Abey Smallcombe
01647 281282
jill@abeysmallcombe.com
www.abeysmallcombe.com

Back to Earth (Crediton, Devon)
Chris Brookman
01363 866999 backearth@aol.com
www.zedland.co.uk/backearth

**Cob in Cornwall Ltd.
(Manaccan, Cornwall)**
Adam Weismann and Katy Bryce
01326 231 773
info@cobincornwall.com
www.cobincornwall.com

**Cob Construction Company
(Exminster, Devon)**
01392 834969
sue@cobconstructioncompany.com
www.cobconstructioncompany.com

Earthed (Sheffield)
Annabel and Alan
www.earthedworld.co.uk

**Kevin McCabe
(Ottery St Mary, Devon)**
01404 814 270
kevin.mccabe2@btopenworld.com
www.buildsomethingbeautiful.com

J & J Sharpe (Okehampton, Devon)
01805 603587
mail@jjsharpe.co.uk
www.jjsharpe.co.uk

Mike Wye (Devon)
01409 281644
sales@mikewye.co.uk
www.mikewye.co.uk

**Seven Generations
Natural Builders (USA)**
Tim Rieth, Andy and Sasha
001 503 886 9923
tim@sgnb.com
www.sgnb.com

ARCHITECTS & APPROVED BUILDING INSPECTORS

(RIBA=Royal Institute of British
Architects) *Designers who have
experience with cob buildings*

Arco 2 (Bodmin, Cornwall) (RIBA)
01208 832990
ianarmstrong@arco2.co.uk
www.arco2.co.uk

**Eco Arc Architecture (North
Yorkshire) (RIBA)**
Andrew Yeats
01904 468752
ecoarc@ecoarc.co.uk
www.ecoarc.co.uk

**Matt Robinson
(St. Martin, Cornwall)**
01326 221339
carvallack@aol.com

**Paul Bedford
(Totnes, Devon) (RIBA)**
Bedford and Jobson Architects
01803 840333
paulbedfordriba@btinternet.com
www.architecturedevon.co.uk

**Associated Architects
(Birmingham) (RIBA)**
0121 233 6600
mail@associated-architects.co.uk
www.associated-architects.co.uk

**Rodda Lloyd Travers Architects
(Penzance, Cornwall) (RIBA)**
01736 367646
admin@rltarchitects.co.uk
www.rltarchitects.co.uk

**PDP Green Consulting Ltd
(Camborne, Cornwall) (RIBA)**
Nick Donaldson
01209 614920
reception@pdpgreen.co.uk

**Approved independent
building inspectors**
jhai Ltd.
Jon Hollely
01308 488 656
jha@approvedinspector.com
www.approvedinspector.com

ORGANISATIONS

The Association for Environment
Conscious Building (AECB)
01559 370908
info@aecb.net
www.aecb.net

British Reed Growers Association
01603 629871

The Building Limes Forum
admin@buildinglimesforum.co.uk

Cal-Earth-California
Institute of Earth Art
and Architecture (USA)
001 760 224 0614
www.calearth.org

The Canelo Project (USA)
Bill and Athena Steen
001 520 455 5548
www.caneloproject.com
*Publications and workshops on straw-
bale and natural building.*

Centre for Alternative
Technology (CAT)
01654 702 400
info@cat.org.uk
www.cat.org.uk
*Educational courses in sustainable
building construction and design.
Supplier of 'green' literature.*

Centre for Earthen Architecture
Plymouth School of Architecture
01752 233630

Cob Cottage Company (USA)
001 541 942 2005
www.deatech.com/cobcottage/
*North American School of Building.
Training in cob and natural building.*

Cornwall Sustainable
Building Trust
01726 68654
www.csbt.org.uk
Courses on traditional building.

CRATerre-EAG (France)
0033 74 96 60 56
www.craterre.archi.fr
*Professional training in earth
construction, architecture
and engineering.*

Devon and Cornwall Master
Thatchers Association
01460 234 477

Devon Earth Building
Association (DEBA)
Devon County Council
Peter Childs
South Coombe, Cheriton Fitzpaine,
Crediton, Devon, EX17 4PH
01363 866 813
*Annual newsletter and pamphlets on
cob building.*

East Anglian Earth Buildings
Group (EARTHA)
01953 601701

English Heritage
0207 973300

Fox Maple School of
Traditional Building (USA)
001 207 935 3720
foxmaple@foxmaple.com
www.foxmaple.com

Heritage Council Ireland
056 7770788
heritage@heritage.iol.ie
www.heritagecouncil.ie

Historic Scotland
0131 668 8600

Institute of Historic
Building Conservation
01892 618 323

Mt Pleasant Eco Park (Pioneer
Environmental Building Company
Ltd) Porthtowan, Cornwall
Tim Stirrup
01209 891 500
enquiries@pioneerebc.co.uk
www.pioneerebc.co.uk

National Society of
Master Thatchers
01494 443198

National Soils Resources Institute
Cranfield University
01525 863 266

Scottish Ecological
Design Association
01361 840230
www.seda2.org

Society for the Protection of
Ancient Buildings (SPAB)
0207 3771644

Thatch: a great website
www.thatch.org

University of Bath
Department of Architecture
and Civil Engineering
Pete Walker
01225 8266
p.walker@bath.ac.uk
www.bath.ac.uk

University College London (UCL)
Institute of Archaeology
Tim Williams, Louise Cooke,
Sjoerd van der Linde
0207 679 4722
Archaeological focus on earth buildings worldwide – conservation and education

University of Kassel
Building Research Institute
+49 561 88 30 50

Walter Segal Self-Build Trust
01668 213 544
www.segalselfbuild.co.uk

Weald & Downland
Open Air Museum
01243 811464
courses@wealddown.co.uk
www.wealddown.co.uk
Courses on traditional building and examples of vernacular architecture.

Yestermorrow Design/
Build School (USA)
001 802 496 5545
ymshcool@aol.com

MAGAZINES & JOURNALS

Permaculture Magazine
www.permaculture.co.uk

Buildng For a Future magazine
www.newbuilder.co.uk

Devon Earth Building Association
Devon County Council
01363 8668 13
Annual newsletter and pamphlets on cob building.

Cobweb
Bi-annual newsletter by the
Cob Cottage Company,
Oregon USA
001 541 942 2005

Devon Historic Buildings Trust
Devon County Council
01392 382261
Pamphlets on sensitive restoration techniques for cob buildings.

The Last Straw (USA)
www.strawhomes.com
Quarterly newsletter which covers straw-bale building but also has excellent articles on natural building in general, such as earth plastering etc.

BOOK PUBLISHERS & SUPPLIERS

Green Books (Devon)
www.greenbooks.co.uk

Eco-logic Books (Bristol)
www.eco-logicbooks.com

Earth Repair Catalogue
www.permaculture.co.uk

Walnut Books (Ireland)
www.walnutbooks.co.uk

Straw Bale Central
www.strawbalecentral.com

GreenBuilder
www.greenbuilder.com/dawn

The Green Shop
www.greenshop.co.uk

New Builder Bookshop
www.newbuilder.co.uk/bookshop

Dirt Cheap Builder (USA)
www.dirtcheapbuilder.com

BIBLIOGRAPHY

BOOKS

Adobe, Remodeling
and Fireplaces (1986)
by Myrtle Steadman
Sunstone Press
ISBN 0 86534 086 2

An Architecture for People:
The Complete Works of
Hassan Fathy (1997)
by James Steele, Thames & Hudson
ISBN 0 8230 0226 8

The Art of Natural Building (2002)
by Joseph F. Kennedy, Michael G.
Smith and Catherine Wanek
New Society Publishers
ISBN 0 86571 433 9

The Beauty of Straw
Bale Homes (2000)
by Bill and Athena Steen
Chelsea Green Publishing Company
ISBN 1 890132 77 2

Building Green: A Guide to
Using Plants on Roofs,
Walls and Pavements (2003)
by Jacklyn Johnston and John
Newton, London Ecology Unit
ISBN 1 871045 185

Building With Straw Bales:
A practical guide for the
UK and Ireland (2003)
by Barbara Jones, Green Books
ISBN 1 903998 13 1

Building a Wood-Fired Oven
for Bread and Pizza (2003)
by Tom Jaine, Prospect Books
ISBN 090732570X

Building With Earth:
A Handbook (2001)
by John Norton
ITDG Publishing
ISBN 1853393371

Building With Lime,
A Practical Introduction (1997)
by Stafford Holmes and Michael
Wingate, Intermediate Technology
Publications
ISBN 1 853393 84 3

Build Your Own Earth Oven (2001)
by Kiko Denzer
Hand Print Press
ISBN 0 9679846 0 2

Build Your Own Home (2004)
by Tony Booth and Mike Dyson
How To Books Ltd
ISBN 1 85703 901 7

Built by Hand, Vernacular
Buildings Around
the World (2003)
by Bill Steen and Athena Steen
Gibbs Smith
ISBN 1 58685 237 X

Caring For Old Houses (2002)
by Pamela Cunnington
Marston House
ISBN 1899296174

Ceramic Houses and Earth
Architecture: How to
Build Your Own (2000)
by Nader Khalili
Cal-Earth Press
ISBN 1 889625 01 9

Clay and Cob
Buildings (2004)
by John McCann
Shire Publications Ltd
ISBN 0 747805 79 2

The Cobber's Companion,
How To Build Your Own
Earthen Home (1998)
by Michael G. Smith
A Cob Cottage Company Publication
ISBN 0 9663738 0 4

Cob Buildings,
A practical Guide, (2004)
by Jane Schofield and Jill
Smallcombe, Black Dog Press
ISBN 0 952434 15 6

Cottage Building in Cob,
Pise, Chalk and Clay (1913)
by Clough-Williams Ellis
Country Life

The Charm of the
English Village (1985)
by P. H. Ditchfield
Bracken Books
ISBN 0 946495 29 7

Colour, Making and Using
Dyes and Pigments (2002)
by Francois Delamare
and Bernard Guineau
Thames and Hudson
ISBN 0 500301 102 6

Discovering Your Old House: How
to Trace the History of Your
Home (1997)
by David Iredale and John Barrett
Shire Publications
ISBN 0747801436

Earth Building and the Cob
Revival: a Reader (1996)
by The Cob Cottage Company
Cob Cottage Company
Publications

Earth Building: Methods
and Materials, Repair and
Conservation (2005)
by Laurence Keefe.
Taylor and Francis
ISBN 0 415323 22 3

Earth: The Conservation
and Repair of Bowhill, Exeter:
Working with Cob (1999)
by Ray Harrison
English Heritage / James and James
ISBN 1873936648

Earth Construction:
A Comprehensive Guide (2003)
by Hugo Houben
and Hubert Guillard
ITDG Publishing
ISBN 1 853391 93 X

Earth Structures and Construction
in Scotland: A Guide to the
Recognition and Conservation of
Earth Technology in Scotland (1996)
by Bruce Walker and Christopher
McGregor, Historic Scotland
ISBN 1 900168 22 7

Earth to Spirit: In Search
of Natural Architecture (1994)
by David Pearson, Gaia Books
ISBN 0 811807 02 9

Ecohouse 2:
A Design Guide (2005)
by Sue Roaf
Architectural Press
ISBN 0 750657 34 0

The Ecology of Building
Materials (2003)
by Bjorn Berge
Architectural Press
ISBN 0 75606 3394 8

English Bread and
Yeast Cookery (1977)
by Elizabeth David
Penguin Books
ISBN 0 71391 026 7

The Forgotten Art of Building
a Good Fireplace (1974)
by Vrest Orton
Alan C Hood & Co, Inc.
ISBN 0 91146 917 6

Green Building Handbook,
Volume One (1997)
by Woolley, Kimmins,
Harrison and Harrison
SPON Press
ISBN 0 41922 690 7

The Green Building Bible,
2nd Edition (2005)
by Keith Hall (editor)
Green Building Press
ISBN 1 898130 02 7

The Hand-Sculpted House:
A Practical and Philosophical
Guide to Building a
Cob Cottage (2002)
by Ianto Evans, Michael
G. Smith & Linda Smiley
Chelsea Green Publishing
Company
ISBN 1 890132 34 9

Home-made Homes: Dwellings
of the Rural Poor in Wales (1988)
by Eurwyn William
The National Museum of Wales
ISBN 0 720003 20 2

Home Work:
Hand Built Shelter (2004)
by Lloyd Kahn, Shelter
ISBN 0 936070 33 1

Introduction to
Permaculture (2000)
by Bill Mollison with
Reny Mia Slay, Tagari Publications
ISBN 0 908228 08 2

Lime In Building:
A Practical Guide (1997)
by Jane Schofield, Black Dog Press
ISBN 0 952434 12 1

Living Under Thatch (2004)
by Barry O'Reilly, Mercier Press
ISBN 1 856354 29 6

The Mud Wall in England (1984)
by J. Harrison
Ancient Monuments Society

The Natural House (2000)
by Daniel D. Chiras
Chelsea Green Publishing Company
ISBN 1 890132 57 8
The Natural Paint Book: The
complete guide to natural paints,
recipes and finishes (2002)
by Lynn Edwards and Julia Lawless
Kyle Cathy Ltd
ISBN 1 856264 32 7

The Natural Plaster Book:
Earth, Lime and Gypsum Plasters
for Natural Homes (2003)
by Cedar Rose Guelberth
and Dan Chiras
New Society Publishers
ISBN 0 086571 449 5

The Nature of Order: An Essay on the Art of Building and The Nature of the Universe (2005)
by Christopher Alexander
Centre for Environmental Structure
ISBN 0 972652 93 0

The New Natural House Book (1998)
by David Pearson
Simon and Schuster
ISBN 0 68484 733 7

Out of Earth II: National Conference on Earth Buildings (1995)
by Linda Watson & Rex Harries (editors), University of Plymouth
ISBN 0 905227 40 9

Places of the Soul (1990)
by Christopher Day
The Aquarian Press
ISBN 0 850308 80 1

Rocket Stoves (2004)
by Ianto Evans
Cob Cottage Company

Shelter (1973)
by Lloyd Kahn
Shelter Publications
ISBN 0 936070 11 0

A Shelter Sketchbook (1997)
by John S. Taylor
Chelsea Green Publishing Company
ISBN 1 890132 02 0

The Spell of the Sensuous (1996)
by David Abram
Vintage Books
ISBN 0 67977 639 7

The Stonebuilder's Primer (2003)
by Charles Long
Firefly Books
ISBN 1 552092 98 4

The Straw Bale House (1994)
by Athena Steen, Bill Steen, David Bainbridge with David Eisenberg
Chelsea Green Publishing Company
ISBN 0 930031 71 7

Terra Britannica (2000)
by John Hurd and Ben Gourley (editors), James and James
ISBN 1 902916 13 1

Thatch Volume 6 (2000)
by Jo Cox and John Letts
James and James Ltd
ISBN 1 873936 96 6

Village Buildings of Britain (1991)
by Matthew Rice, Time Warner
ISBN 0 31672 625 7

The Whole House Book (2005)
by Cindy Harris and Pat Borer
Centre for Alternative Technology Publications
ISBN 1 902175 22 0

World Soils (1990)
by E. M. Bridges
Cambridge University Press
ISBN 0 52129 339 1

Your Country Cottage: A Guide to Purchase and Restoration (1970)
by Robert Edmunds
David and Charles

PAMPHLETS, MAGAZINES & JOURNALS

Appropriate Plasters, Renders and Finishes for Cob and Random Stone Walls in Devon (2nd ed.)
by Paul Bedford, Bruce and Liz Induni and Larry Keefe
Devon Earth Building Association

Building Regulations and Historic Buildings (2004)
by English Heritage

Cob and the 1991 Building Regulations (March 1997)
by Tony Ley and Mervyn Widgery
The Devon Earth Building Association

The Cob Buildings of Devon: History, Building Methods and Conservation
Volume 1, 1992
by The Devon Historic Building Trust

The Cob Buildings of Devon, Repair and Maintenance
Volume 2, 1993
by Larry Keefe
The Devon Historic Buildings Trust

'Green Roofs' in Building For A Future Magazine Volume 15, no.1, Summer 2005. pp.28-30

The Last Straw Journal
Issue no.33, Spring 2001

The Last Straw Journal
Issue no.38, Summer 2002

The Last Straw Journal
Issue no.43, Fall 2003.

Thatch In Devon (March 2003)
by Devon County Council

Thatch and Thatching: A guidance note (2000)
by English Heritage

Index